Miguel Angel Gardetti ·
Subramanian Senthilkannan Muthu
Editors

The UN Sustainable Development Goals for the Textile and Fashion Industry

Springer

Editors
Miguel Angel Gardetti
Center for Studies on Sustainable Luxury
Buenos Aires, Argentina

Subramanian Senthilkannan Muthu
SgT Group & API
Hong Kong, Kowloon, Hong Kong

ISSN 2197-9863 ISSN 2197-9871 (electronic)
Textile Science and Clothing Technology
ISBN 978-981-13-8786-9 ISBN 978-981-13-8787-6 (eBook)
https://doi.org/10.1007/978-981-13-8787-6

This Springer imprint is published by the registered company Springer Nature Singapore Pte Ltd.
The registered company address is: 152 Beach Road, #21-01/04 Gateway East, Singapore 189721, Singapore

Textile Science and Clothing Technology

Series Editor

Subramanian Senthilkannan Muthu, SgT Group & API, Hong Kong, Kowloon, Hong Kong

This series aims to broadly cover all the aspects related to textiles science and technology and clothing science and technology. Below are the areas fall under the aims and scope of this series, but not limited to: Production and properties of various natural and synthetic fibres; Production and properties of different yarns, fabrics and apparels; Manufacturing aspects of textiles and clothing; Modelling and Simulation aspects related to textiles and clothing; Production and properties of Nonwovens; Evaluation/testing of various properties of textiles and clothing products; Supply chain management of textiles and clothing; Aspects related to Clothing Science such as comfort; Functional aspects and evaluation of textiles; Textile biomaterials and bioengineering; Nano, micro, smart, sport and intelligent textiles; Various aspects of industrial and technical applications of textiles and clothing; Apparel manufacturing and engineering; New developments and applications pertaining to textiles and clothing materials and their manufacturing methods; Textile design aspects; Sustainable fashion and textiles; Green Textiles and Eco-Fashion; Sustainability aspects of textiles and clothing; Environmental assessments of textiles and clothing supply chain; Green Composites; Sustainable Luxury and Sustainable Consumption; Waste Management in Textiles; Sustainability Standards and Green labels; Social and Economic Sustainability of Textiles and Clothing.

More information about this series at http://www.springer.com/series/13111

Preface

In 2012, Rio de Janeiro Summit, organised by the United Nations and called "Rio +20", analysed the progress made since the summit held in Rio de Janeiro in 1992, and it also announced that the Millennium Goals were to be replaced—starting in 2015—by the Sustainable Development Goals, also called "the 2030 Agenda for Sustainable Development Goals".

In September 2015, the United Nations General Assembly approved the agenda, which sets up a transformational view to economic, social and environmental sustainability. The current development paradigm should be transformed into an inclusive paradigm based on sustainable development and with a long-term vision. The Agenda comprises the 17 goals below, which, in turn, include 169 targets:

Goal 1: End poverty in all its forms everywhere.
Goal 2: Zero hunger. End hunger, achieve food security and improved nutrition and promote sustainable agriculture.
Goal 3: Good Health. Ensure healthy lives and promote well-being for all at all ages.
Goal 4: Quality education. Ensure inclusive and equitable quality education and promote lifelong learning opportunities for all.
Goal 5: Gender equality. Achieve gender equality and empower all women and girls.
Goal 6: Clean water and sanitation. Ensure availability and sustainable management of water and sanitation for all.
Goal 7: Affordable and clean energy. Ensure access to affordable, reliable, sustainable and modern energy for all.
Goal 8: Decent work and economic growth. Promote sustained, inclusive and sustainable economic growth, full and productive employment and decent work for all.
Goal 9: Industry, innovation, infrastructure. Build resilient infrastructure, promote inclusive and sustainable industrialisation and foster innovation.
Goal 10: Reduced inequalities. Reduce inequality within and among countries.

Goal 11: Sustainable cities and communities. Make cities and human settlements inclusive, safe, resilient and sustainable.
Goal 12: Responsible consumption and production. Ensure sustainable consumption and production patterns.
Goal 13: Climate Action. Take urgent action to combat climate change and its impacts.
Goal 14: Life below water. Conserve and sustainably use the oceans, seas and marine resources for sustainable development.
Goal 15: Life on land. Protect, restore and promote sustainable use of terrestrial ecosystems, sustainably manage forests, combat desertification and halt and reverse land degradation and halt biodiversity loss.
Goal 16: Peace, justice and strong institutions. Promote peaceful and inclusive societies for sustainable development, provide access to justice for all and build effective, accountable and inclusive institutions at all levels.
Goal 17: Partnership for the goals. Strengthen the means of implementation and revitalise the global partnership for sustainable development.

This book presents four chapters that relate textile and fashion to some SDGs. For example, the work titled "Traceability & Transparency: A Way Forward for SDG 12 in the Textile and Clothing Industry" developed by Natalia Papú Carrone presents a deepened understanding of what traceability and transparency concerns are all about, and how they can constitute an enabler to accelerate the industry efforts towards achieving SDG 12. Then, "Sustainable Development Goal 12 and Its Relationship with the Textile Industry" developed by Marisa Gabriel and María Lourdes Delgado Luque analyses SDG 12 and how it can be applied to the textile industry, considering the circular economy as a way towards. The Sanjoy Debnath´s work titled "Flax Fibre Extraction to Fashion Products Leading Towards Sustainable Goals" covers, to some extent, cultivation and extraction of fibre and further processing into yarn and fabric up to fashion garments. It also touches on aspects such as reuse and bio-disposal. This chapter further analyses the UN Sustainable Development Goals in the flax value chain. Finally, Radhakrishnan Shanthi in his Chapter "Sustainable Consumption and Production Patterns in Fashion" analyses the role of sustainable design development, the awareness of slow fashion and change in consumer mindset to attain Goal 12 (Responsible Consumption and Production), UN Sustainable Development Goals.

Buenos Aires, Argentina — Miguel Angel Gardetti
Hong Kong — Subramanian Senthilkannan Muthu

Contents

Traceability and Transparency: A Way Forward for SDG 12 in the Textile and Clothing Industry

Natalia Papú Carrone

Abstract Sustainable Development Goal, SDG, 12 calls for a profound business transformation towards sustainable consumption and production patterns. This involves the entire value chain from a holistic perspective, from raw material to consumer, both globally and locally. Target 12.8 specifically identifies the need for people everywhere to '*have the relevant information and awareness for sustainable development and lifestyles in harmony with nature*'. To enable this shift, industry practitioners and academics have recognised traceability as the necessary first step for informed decision-making. Traceability, according to the International Organization for Standardization, ISO, refers to the '*ability to identify and trace the history, distribution, location, and application of products, parts, materials, and services*'. Full implementation of traceability systems will allow industry partners to have access to reliable, comprehensive data of their business activities as well as their related environmental and social impact. Once this information is traced and available for firms, transparency will enable all stakeholders to have access to the relevant information needed to make informed decisions, including but not limited to customers and business partners. Both transparency and traceability support visibility throughout the textile and clothing value chains and therefore contribute to build trust between stakeholders. Overall, the purpose of this chapter is to present a deepened understanding of what traceability and transparency concern, and how they can constitute an enabler to accelerate industry's efforts towards achieving SDG 12. Available literature is reviewed thoroughly and supported by examples of implemented industry practices.

Keywords Traceability · Transparency · Textiles and clothing · T&C · Sustainable development

N. Papú Carrone (✉)
Department of Business Administration and Textile Management, University of Borås, 501 90 Borås, Sweden
e-mail: natipapu@gmail.com

M. A. Gardetti and S. S. Muthu (eds.), *The UN Sustainable Development Goals for the Textile and Fashion Industry*, Textile Science and Clothing Technology,
https://doi.org/10.1007/978-981-13-8787-6_1

1 Introduction

Textiles and clothing, T&C, and its related industries have a significant impact on the environmental and social footprints on our planet, mainly driven by resource—and labour-intensive practices [26]. Manufactured consumables, such as T&C, together with mobility, are the social needs driving the largest carbon footprint throughout their value chains [8]. Furthermore, from the materials being extracted, sourced and placed into production, most of them are used only once, being disposed within the first year of use (ibid.). The current linear system in which a business operates is based on a take-make-use-dispose model, which derives from the assumption that there is an endless availability of easily accessible and high-quality fossil fuels [58].

Indeed, this is not the case for our planet, and Sustainable Development Goal, SDG, 12 calls for a profound transformation in the ways current businesses operate. SDG 12 is aimed at achieving more sustainable consumption and production patterns [48]. This involves the entire value chain from a holistic perspective, from raw material to consumer, both globally and locally. Fletcher [17] suggests that this transformation has to be supported by a mindful ideology from all involved stakeholders, which in turn, fosters the identification of the underlying values and economic drivers of fashion production. On the same regard, Joy and Peña [26] postulate the need to adopt 'slower' fashion which focuses on longevity and artisanship, in contrast to 'fast' fashion's accelerated consumption loop, where homogeneity and short-term satisfaction are exacerbated.

Behind the scenes of consumption practices, lay opaque production operations. Globalised and complex value chains scattered around the world characterise the current operating status of the T&C environment [9, 50]. The number of suppliers involved in each value chain has grown exponentially during the last decades, resulting in intricate value networks with a low degree of control [13]. Thus, transparency and trust have become more difficult to acquire in current operating practices and are becoming a real concern for the T&C industry [1]. Extremes have been encountered when deadly incidents such as the Rana Plaza collapse in 2013 or the Ali Enterprises factory fire in 2012 left thousands of dead and injured and pushed the industry to open its eyes [9]. The public pressure of NGOs and activist groups has been key to pursue a change in this regard.

Within SDG 12, target 12.8 specifically identifies the need for people everywhere to '*have the relevant information and awareness for sustainable development and lifestyles in harmony with nature*' [48]. To enable this shift, industry practitioners and academics have recognised traceability as the necessary first step for informed decision-making [19]. Traceability systems, therefore, offer a possible solution to track and trace activities of each actor in the value chain, hence, optimising processes and control throughout the network, as well as enabling the consumer-facing brands to validate their sustainability claims, enhancing reputation and securing the fight against counterfeit [29]. Traceability, coupled with public transparency, also enables consumers to access relevant information about a brand's social and environmental practices, giving consumers with purchasing power an opportunity for more informed decision-making [9]. Moreover, both industry and consumers must be backed by government policies which develop a holistic approach towards the industry [40]. Therefore, local and national governments as well as international organisations play an extremely important role in securing transparency and traceability in the T&C sector. Being consistent with internationally recognised guidelines such as the UN Guiding Principles on Business and Human Rights, the ILO International Labour Standards or the OECD Due Diligence Guidelines for Responsible Business Conduct seems to be the right path to be on [9].

Traceability and transparency therefore allow both industry partners as well as consumers to have access to reliable, comprehensive data of business activities as well as their related environmental and social impact. Gaining visibility into the realities of the different processes enables the improvement of work ethics as well as enabling work towards more responsible business conduct [9, 26]. Additionally, traceability can assist the advancement of product quality and adequate delivery times [29], as well as the handling of specific data relevant for product safety, such as the use of chemicals in dyeing or finishing processes [56]. All of these issues not only upgrade the industry's efficiency and effectiveness but are a prerequisite in changing our current linear business operations to a circular model [19].

Currently, the trend of businesses adopting traceability and transparency practices is rising rapidly [9, 19]. Nevertheless, the scale of adoption is still insufficient to level the playing field in the industry. Many businesses started publishing supplier lists involving names and addresses, mainly in response to NGOs, activist groups or consumer campaigns [7, 27, 34] or due to local or regional regulations [5]. Although most of this disclosure is motivated by external pressures, it remains a robust mechanism to foster corporate accountability and initiate a path towards achieving SDG 12.

Overall, the purpose of this chapter is to present a deepened understanding of what traceability and transparency concern, and how they can constitute an enabler to accelerate industry's efforts towards achieving SDG 12. Available literature was reviewed thoroughly and supported by examples of implemented industry practices. Following this introduction, the concepts of Sustainable Development, Traceability and Transparency will be presented, supported by an introduction to the most recurrent traceability and transparency schemes present in the T&C industry. The main issues whilst implementing these practices are discussed at the end of the corresponding sections.

2 Sustainable Development Goal 12

The United Nations, UN, through its Agenda 2030 framework has the purpose of achieving sustainable development for all at a global scale. Sustainable development as initially defined in the Brundtland Commission Report in 1987 expects to '*meet the needs of the present without compromising the ability to meet future generations' needs*' [6]. The UN Agenda 2030 discusses the urgency to act towards this goal, therefore strengthening universal peace and freedom and eradicating poverty [49].

Poverty and inequality have not only a major impact on social living standards but also on environmental distress [36]. The massive industrialisation in the mid-1900s, as well as the population increase and the rise of the middle class worldwide, has put enormous pressure on natural reserves and the environment whilst consuming our resources in a take-make-use-dispose, linear, wasteful manner [8, 42]. If we are to continue with current consumption and production patterns, it is agreed that the essential planetary resources will be depleted before replenishment can take place [37].

Shifting value chain and consumption practices, as well as enabling technological changes in current production methods, could therefore minimise the negative impact on the environment, as well as raising living standards [36]. Sustainable development has also become a focus for the private sector. The balance is being sought between the social, economic and environmental dimensions of the management of businesses [4]. Companies are increasingly measuring their impacts and identifying areas of improvement to make better and more informed decisions [23]. This awareness does not always come from the inside of the company, and it is customer demands, governments and activist groups which may pressure companies to adopt more sustainable practices in their supply chains [36, 42]. Hence, many supply chain actors are revising their ecological footprint in order to reduce it, to remove toxic chemicals and reduce the carbon footprint throughout their value chain processes [23]. Textile production has a major impact within pollutants discharge in air and water and energy and water consumption and inefficient processes exacerbate this matter [56]. Therefore, improvement of the ecological footprint is mainly sought through energy efficiency, forecasting accuracy improvement or process restructuring [23]. However, it is of significance to comprehend that the approach to sustainable development should be systemic and processes cannot be observed and analysed in isolation.

This is the reason why SDG 12 aims at a profound business transformation holistically analysing and acting upon consumption and production patterns. Focused around the basis of decoupling economic growth from resource use as the way forward towards more sustainable consumption and production practices [48], SDG 12 is aligned with the shift from a linear economic and business model to a circular one. The targets within this SDG expect the uptake of relevant country-level policies and programmes on sustainable consumption and production, the reduction in global food waste by half and the achievement of the sustainable management of natural resources, chemicals and emissions by 2030. It also encourages the adoption of sustainable practices by companies and public procurement. Cross-cutting all of these issues, the Agenda 2030 expects to motivate support to developing countries to strengthen capabilities and technology, as well as to promote local cultures and sustainable tourism, and rationalise the subsidies for fossil fuels, which currently promote a linear take-make-dispose economy.

In regard to traceability and transparency, SDG 12 postulates target 12.8, which specifically identifies the need for people everywhere to '*have the relevant information and awareness for sustainable development and lifestyles in harmony with nature*' [48]. As all targets are interrelated between each other, 12.8 remains an enabler for the successful achievement of other SDG targets. Traceability and transparency, therefore, are essential tools for the SDGs [30]. They also entail the adoption of digital technology as a key to 'track and optimise resource use' whilst building stronger linkages between value chain partners. The incorporation of digital technology, according to the Platform for Accelerating the Circular Economy, PACE, is one of the seven key elements that facilitate the circular economy [8]. On this target, the UN has reported that during 2018 the number of national policies and initiatives on sustainable consumption and production, as well the number of companies reporting on sustainable practices, has increased compared to previous years [48]. Nevertheless, much remains to be done. Kumar et al. [30] suggest that sustainability cannot be claimed without the relevant traceable information which backs it, leading to the practice of 'greenwashing' where a sustainability claim is made but no data on the subject can be retrieved to substantiate that claim.

Next, we will dive into the specifics of traceability and transparency practices. To access further information on SDG 12 and the Agenda 2030, a comprehensive overview can be found at the UN Sustainable Development website.[1]

3 Setting a Common Understanding of Traceability and Transparency

Traceability, according to the International Organization for Standardization, ISO, refers to the 'ability to identify and trace the history, distribution, location, and application of products, parts, materials, and services' [25]. Full implementation of traceability systems could therefore allow industry partners to have access to reliable, comprehensive data of their business activities as well as their related environmental and social impact. Voluntary traceability schemes and technologies as well as standard measurement tools are available to gain deeper insight into how products are made [19]. This could not only substantiate environmental and ethical transgression claims but could also promote more efficient management of value chains, as well as give access to the data needed for businesses to communicate accurately the impact of their activities (ibid.).

Once this data is traced and available for firms, transparency can enable all stakeholders to have access to the relevant information needed to make informed decisions, including but not limited to customers and business partners. Transparency, consequently, relates to the disclosure of information [12]. This concept overarches both having visibility of the value chain as well as taking action and managing risks more effectively due to the visibility gained [32]. It enables stakeholders which are external to the production process to gain knowledge about it [12]. It also allows value chain partners to have a clear understanding of what happens within partner companies. Gaining this knowledge has several benefits, although it also may create interest and power conflicts between supply chain partners. This is why it remains important to

[1]UN Agenda 2030 for Sustainable Development https://sustainabledevelopment.un.org.

work towards transparency in the value chain in a collective way between stakeholders and not only demanded as a requirement by one business partner (ibid.).

Transparency also relates to reporting. It has been postulated that reporting procedures need to be more '*clear, consistent and comparable*' [46]. This may improve legitimacy and verify sustainability claims made by businesses [12]. Some research shows that companies' sustainability reporting may be challenged in its credibility due to the reporting done selectively, disregarding the areas of the impact the company does not wish to show [36]. Hence, there is a need for an integrated and comparable mechanism.

As mentioned previously, target 12.8 of Sustainable Development Goal 12 of the Agenda 2030 specifically identifies the need for people everywhere to '*have the relevant information and awareness for sustainable development (…)*' [48]. Both transparency and traceability support visibility throughout the textile value chains and therefore could contribute to build trust between stakeholders [36]. Currently, an increasing amount of businesses, including brands and manufacturers, implements standard measurement tools and traceability schemes which increase the availability of comparable data out there to benchmark corporate sustainable practices [19]. A few businesses are also testing and implementing new technologies such as blockchain-based systems. Some of the practices and schemes of T&C traceability and transparency are presented in the sections below, discussing the main benefits and challenges of their implementation.

4 T&C Traceability

4.1 Overview

There are several definitions of the concept of traceability. The ISO defines traceability as the 'ability to identify and trace the history, distribution, location, and application of products, parts, materials, and services' [25]. This is the most commonly used definition. Global Standards One, GS1, a global not-for-profit organisation that sets and implements standards for value chains adds that together with the tracing backward, that ISO distinguishes, traceability also includes the '*ability to track forward the movement through specified stage/s of the extended supply chain*' [22]. Macchion et al. [34] support this categorisation of traceability into the tracing (backwards) and the tracking (forward) practices. Further, the United Nations Economic Commission for Europe, UNECE, adds that traceability can be understood as '*a method to substantiate a claim (…) relating to a product, service or business process based on available information*' [55]. This concept definition can be easily related to the fact that SDG 12.8 requires everyone to have access to the relevant information and awareness, and this entails that this relevant information needs to be substantiated with accurate data, to avoid misrepresentation of facts or other unethical practices such as 'greenwashing'. As consumers' and producers' interest grows in gaining insight into a product's production processes and its impact [55], the combination of the three definitions stated above acquires relevance. The quality of the data emerging from traceability systems should also be an area of focus, as usual data on the sourcing and the use of materials is widely available, but data on the end-of-use and disposal stages of the value chain are deeply uncertain [8].

Implementation of traceability was firstly aimed at logistical purposes. In the mid-1980s, the focus shifted towards ensuring food product safety [26]. In the past decades, the implementation has broadened, and several industries have incorporated traceability systems and practices due to attaining multiple benefits. Firstly, as already discussed, it gives both businesses and consumers the ability to track and to gain awareness of value chain information through data linked to a specific product [13]. Secondly, it fosters collaboration between business partners, by implementing traceability systems which are interoperable and integrated [22]. Therefore, it helps integrate the value chain by extending the reach of the information boundaries of each value chain actor, making them aware of the relevant practices of their partners [30]. Information sharing and value chain data integration also lead to the improvement of communication and control of the business networks [34]. All of the above enable businesses and other stakeholders to manage the risks throughout the value chains in a more effective manner, keeping extant documentation about processes and procedures [24]. Most social risks are related to the respect for human rights, worker's rights, freedom of association, child labour, health and product safety, corruption practices and disloyal competition [56]. Environmental risks mostly refer to overconsumption of energy and water resources, use of harmful chemicals and mismanagement of wastewater and gas emissions discharges (ibid.).

Furthermore, a benefit closely related to the implementation of traceability systems is the possibility to control closely the product quality, the material flow, product defects and logistics such as warehousing and transportation [1, 29]. It also enables a better adjustment of forecasts to market changes and reduces the risks of demand uncertainty by enabling, for example, geo-localisation of customers and shopping behaviour segmentation [34]. These authors suggest as well that traceability may be useful when aiming at innovating at certain production stages. In addition to the benefits mentioned above, a relevant area of implementation to mention is the fight against counterfeited products in the T&C industry [31].

4.2 T&C Traceability Schemes

Traceability systems focus on four main questions: what, when, where and why by tracking identifiers which are related to a specific asset: a material, a component or a product which means to be traced [55]. These identities are unique to each asset and are key to being able to track forward or trace back the data recorded for the asset [29]. In Fig. 1, the material and the information flows of the asset and events involved in a clothing value chain are exemplified.

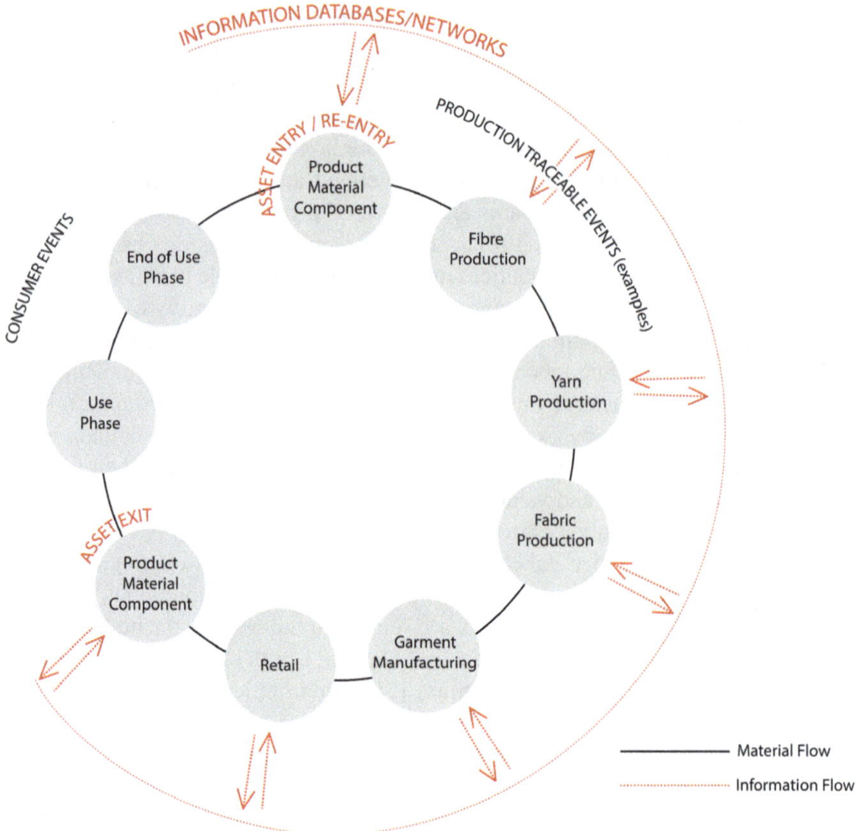

Fig. 1 Proposed overview of a traceable asset and events in T&C value chains

Currently, some regulations expect the tracing of hazardous chemicals throughout production processes to guarantee that final products are safe and free from these. For example, the Registration, Evaluation, Authorisation and Restriction of Chemicals, REACH, regulation in the European Union, EU, was enforced in 2007 and expects to improve human and environmental safety from chemical risks, whilst decreasing the amount of animal testing conducted [11]. In order for this to happen, the EU regulation places the '*burden of proof*' on the businesses, which must identify and manage the chemical substances that are connected to the products they manufacture and market within the EU [11]. In these cases, traceability can be a useful tool to track chemicals and 'guarantee' that your business is in line with the regulations. Although there are not many other mandatory requirements to trace products within T&C value chains, traceability schemes of a voluntary nature are also increasing their uptake [30]. Supporting legislations and standard development to identify the proper traceability scheme design for T&C value chains is extremely necessary to enhance the adoption of these practices within the industry [34]. In this regard, the UNECE together with the International Trade Centre, ITC, for example, are working on the development of UN policy recommendations and standards for a traceability framework for the T&C sector [52]. An internationally recognised standard on textile traceability, which takes into account all the voluntary traceability developments which exist to the date, would tackle many implementation challenges from current traceability schemes related to the development of common semantics and interoperable networks between them. This, in turn, would facilitate information sharing between business partners and from country to country [29].

Regarding traceability schemes, this overview is divided into two parts. Firstly, different types of traceability systems will be presented, such as product segregation, mass balance and book and claim. Later in this section, an overview of existing types of technologies which are used to facilitate traceability practices are introduced, for example, barcodes, RFID tags and blockchain-based systems.

Traceability Systems

There are different ways in which an asset (material, product, component) can be traced throughout the value chain. This is dependent on the way the asset is accounted for, if this is done in an aggregated manner or if each asset is accounted for separately. The most comprehensive and detailed traceability system is known as product segregation. This entails that the certified materials or products for a certain type of production (e.g. organic material) are traced separately from non-certified material [55]. This traceability system can be implemented either by mixing certified materials from different producers to reach larger scales, i.e. in bulk commodities, or by preserving the identity of each certified material throughout the entire value chain, from the first processor to the consumer (ibid.).

A brand which has been working extensively towards reaching full traceability by the end of 2019 is the Swedish brand ASKET. This company was launched in 2015 and has currently achieved 71% of its full traceability goal [3]. The purpose behind fully tracing their garments lays with enhancing people's consciousness, both within business as well as consumers, in order to foster more informed choices, in line with Fletcher [17] mindfulness in fashion. The traceability system this brand is implementing is based on obtaining verified certificates in accordance with their established rules of performance for each subprocess of their value chain. This is coupled with on-site visits and travels, to enhance the knowledge about their own and their partners' operations. Certification bodies and internal and external audits therefore become relevant for traceability systems. The traceable information is fed into a traceability scoring system which organises the value chain processes into four main areas: manufacturing, milling, raw material and trims, approximately tracing 400 subprocesses in their value chain [3]. The information traced is later connected to transparency practices, such as garment labelling or online disclosure, giving the consumer a much more informed overview than the usual '*Made in*' label. These differences can be observed in Figs. 2 and 3. Interestingly, this brand also communicates that they have not reached their goal yet and therefore inform the areas in which they are still in the process of tracing back their production. This practice not only aids trust building

Fig. 2 'Made in' standard label [18]

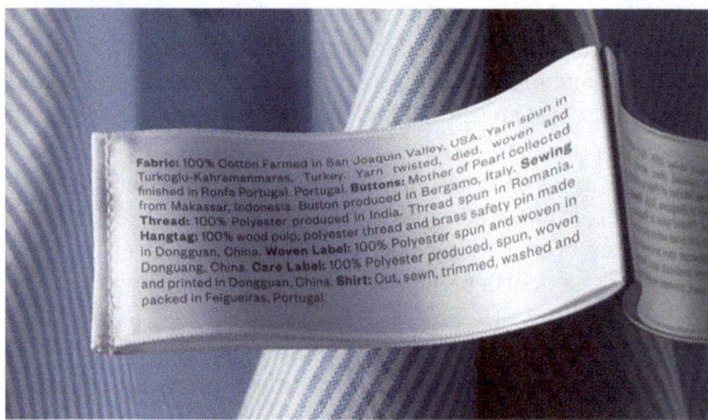

Fig. 3 ASKET's detailed 'Made in' label [16]

with the consumer but could also lead to industry collaborations which accelerate the process of reaching full traceability of the garment's components and materials.

Although product segregation systems provide a detailed understanding of the traceable assets and can identify individual assets in the process, it presents many challenges for smallholders in the value chain regarding technology, scale and business capabilities [55]. Hence, other traceability systems which entail less requirements are currently more widely spread amongst the industry. These systems may be a mass balance, in which the claim (e.g. organic content) is not segregated by asset but is referred to the aggregated amount of certified material available in the assets [51]. An example of this system would be a company that claims that 40% of their cotton is organic, though there is no clear definition on which the certified material is, and which is not, as they travel together throughout the value chain. Another system which may be utilised is Book and Claim, where there is no traceability at each stage of the value chain; however, a company can make a claim on specific processes related to sustainable sourcing (e.g. wastewater treatment, carbon emissions) [55]. This practice could be understood as a first step towards adopting a traceability system.

Traceability Technologies

Several technologies are used in order to track and trace assets throughout value chains. The most commonly used are barcodes, radio-frequency identification tags (RFIDs), magnetic barcodes and other identifiers such as organic chemical marking [29]. Recent research shows new possibilities of technology useful for implementing traceability systems, such as yarn-based tags [29], two-factor product authentication and tracking systems [1] and digital material passports [35]. New developments in blockchain technology also show interesting repercussions for complex and distributed textile value chains, enabling a platform for information sharing whilst guaranteeing the authenticity and validity of the data [13]. In this regard, Sweden-based TrusTrace, a blockchain-based digital collaboration platform that supports product traceability, is an example of these developments [52]. The company has recently joined Fashion for Good's 4th accelerator programme and is expected to sustain significant growth and positioning within the industry by working together with some of Sweden's most well-known brands. All of these technologies constitute different ways to access, share and communicate traceable data, and it remains key to analyse the specific context and objectives of a traceability system to understand which technology provides the best available solution.

4.3 Main Implementation Challenges

Setting aside the benefits and successful uses of traceability systems mentioned previously, throughout available literature and industry practice, a series of challenges emerge when implementing traceability. First of all, setting up a new system, in this case, one that enables tracking and tracing products or processes throughout value chains, is associated to upfront investment costs. Within the textile industry, where a myriad of smallholders integrates the value chain, this may hinder the adoption of traceability practices, especially by micro, small and medium-sized enterprises, MSMEs, and other smallholders in developing or underdeveloped countries [26, 55]. This barrier to adoption leads to the need for collaboration and partnership of all value chain actors in order to split the incentives and create a shared approach to address social and environmental issues [8, 56]. Hence, the need for collaboration places trust between value chain actors at the centre of the stage [26]. Trust also becomes essential for consumer-facing brands due to them being largely reliant on their external manufacturing and raw material suppliers worldwide [30]. Although trust may be uplifted between business partners, there has to be a clear understanding and agreement on the protection of confidential or sensitive data for each partner (ibid.). Moreover, up to the date, there is no single traceability system used worldwide [13], creating a diversity of traceability semantics, its 'language'. The lack of a common traceability language may obstruct and/or delay the flow of information exchange between value chain actors as well as complicate the interoperability between different systems [30].

Other challenges that arise relate mainly to technological aspects of traceability. On one hand, new technological skills and capabilities need to be developed wherever there is a lack of them. An example is the set-up of traceability systems for farmers and field workers, especially when considering upcoming technologies such as those systems running on the blockchain [13]. Nevertheless, some developments have reached simple interfaces through mobile phones, mainly connecting farmers to online marketplaces, though the cost is still high, and it is yet to be spread amongst the value chain [59]. Traceability of livestock at the farms has already adopted in the leather industry when it constitutes a by-product of the food industry [34]. Consequently, relevant knowledge could be derived from these experiences and adapted to the specifics of other value chain stages. Another issue related to technology is the durability of tags (RFID, barcodes, etc) linked to specific products. Agrawal et al. [1] voice that for traceability to ensure security and robustness, tags should endure all stages until reaching the recycling phase at the end-of-use. Covert identifiers which are integrated into textiles and clothing minimise this risk although they present other challenges, as they have a large environmental impact at their end-of-use phase [29].

5 T&C Transparency

5.1 Overview

Following an overview of T&C traceability practices, a more in-depth understanding of schemes that foster transparency is demanded as well, to build complementary practices towards achieving SDG 12. Transparency relates to having access to relevant information, which enables visibility into production and consumption practices [4]. Five main aspects are defined by Vishwanath and Kaufmann [57] as crucial characteristics of transparent information. These are access, comprehensiveness, relevance, quality and reliability. A transparent business behaviour allows consumers to be aware of the social and environmental-related impacts of business operations, therefore motivating them to make more informed choices [4]. Transparency may provide an answer to the accountability question of 'where does my product come from?' [15], empowering other stakeholders involved in the T&C sector, such as consumers, not-for-profit or international organisations, governmental institutions and activist groups, to assess the true impact of business behaviour. Further, transparency is fundamental for business partners along a value chain to grasp the internal practices of others and to build trust throughout transactions and operations.

In this regard, traceability can be seen as a prerequisite for transparency, as it provides assistance in obtaining the necessary information to later be either publicly disclosed or shared between relevant stakeholders [26, 30, 31]. During the last years, it has been made evident that access to relevant and comprehensive information on the raw material production stage of T&C value chains is largely necessary, as a great amount of social and environmental challenges are found at this stage [13]. Nevertheless, the level of information sharing and effectiveness of these practices still remains a challenge due to organisational and technological disparities between value chain partners and geographical, regulatory and cultural differences [30].

Transparency in the field of public disclosure of information informs consumers and the general public about social, environmental and health impacts derived from the textiles and clothing produced and being consumed [20, 26]. Disclosure of comparable and comprehensive information on production practices aids trust building between all stakeholders [9]. Further, it can entitle activist groups, consumer associations and other organisations to sustain claims on human and worker rights violations, pollution and environmental issues and health and safety non-compliance. Transparent disclosure should, however, not be used only as a defamation tool towards brands, but on the other hand, claims found to be true should encourage active work towards promptly finding solutions, introducing businesses as an engine of sustainable corporate citizenship [9, 50]. Working towards industry better practices aligns as well with the businesses' responsibilities to prevent human rights risks throughout their value chains, assigned by the United Nations under the UN Guiding Principles on Business and Human Rights endorsed by the UN Human Rights Council in 2011 [54].

Transparency can also serve the purpose of building brand reputation [9]. Hence, for some businesses it represents a possibility of acquiring an edge of competitive advantage [34]. An improved brand reputation may have a significant impact on the perceived value of a brand, which can also result in an increase in the brand's equity [28].

Achieving increased transparency within a value chain and towards its consumers will also result in a more effective response to the changing demands of the market [38]. This may facilitate the creation of new markets as well [26], adapted to consumer and planetary needs in the midst of a new circular economy.

5.2 T&C Transparency Schemes

Transparency schemes implemented throughout the last decades have mostly been driven and incentivised in three different ways. Firstly, public campaigns pushing brands to disclose value chain information related to social and environmental impacts have been the most common practice. This practice has mainly been driven by activist groups, not-for-profit organisations and consumer associations. Secondly, advancements in this area have also been motivated by new policy developments, mainly at a national or local scale. International policy recommendations have, however, greatly influenced the work of national policy. Lastly, transparency practices

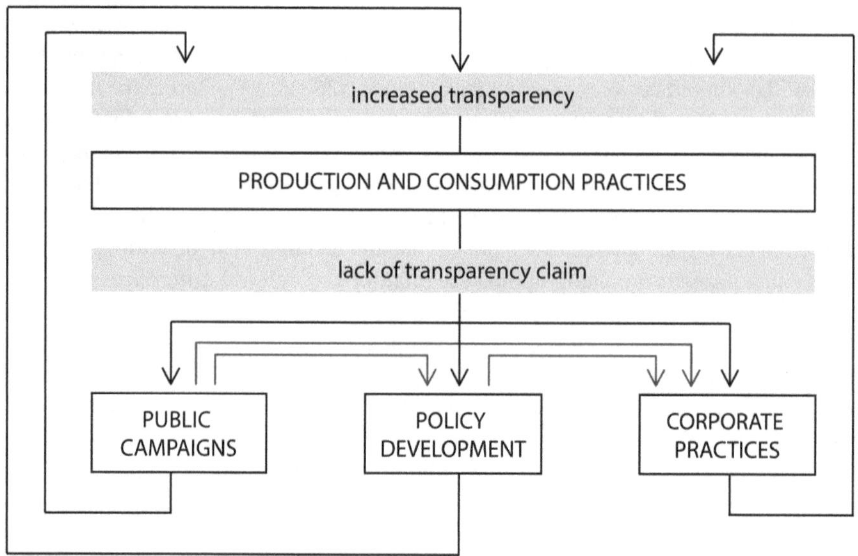

Fig. 4 Proposed overview of transparency practices in T&C

have also been established from the corporate sector, though mainly triggered by public campaign requests. Disclosing supplier lists and adopting voluntary codes of conduct lead transparency practices within the corporate sector, although there have been other developments such as collaborative benchmarking tools in the areas of human rights and value chain sustainability which also have been implemented in the last years (Fig. 4).

Public Campaigns

Until recently, public disclosure of value chain information of production of textile and clothing products was not even considered by businesses [9]. It was said to entail a significant risk towards losing the competitive advantage they had gained by embedding efficiency into their value chains and securing their 'production secret recipe'. The first large disclosure of supplier information was triggered by a campaign led by the United Students Against Sweatshops network, USAS, which demanded businesses which produced US collegiate apparel to disclose their supplier information in the late 1990s (ibid.). Universities in the USA therefore included the disclosure of this information as a clause in their licensing agreements, forcing brands into complying with this request in order to maintain their contracts. Currently, organisations such as the Clean Clothes Campaign, CCC, and Greenpeace continue campaigning and activist work towards achieving effective brands' information disclosure and transparent operations. The Transparency Pledge launched by a group of organisations (CCC, Human Rights Watch, ICAR, IRLF, Maquila Solidarity Network, Worker Rights Consortium, IndustriAll, UNI Global Union and ITUC CSI IGB) after the Rana Plaza disaster in 2013, works towards achieving transparency regarding value chain practices and labour and social standards [9]. A study from Kamal and Deegan [27] shows that from the late 1990s onwards there has been an increasing trend in information disclosure of social and environmental practices in Bangladeshi textile and clothing businesses. Although the trend in most cases is positive, much work remains to be done. The same study argues that disclosure practices in production countries, such as Bangladesh, are mainly driven by the global community's expectations, which therefore supports the need to work collaboratively at a global scale through well-designed information and educational campaigns and programmes to foster the necessary transparent behaviour to achieve SDG 12 targets [41]. On the environmental issues, Greenpeace has led multiple campaigns, whilst the Detox Campaign launched in 2011 was the most relevant yet for the T&C industry. It aimed at eliminating the use and discharge of hazardous chemicals from T&C value chains [21]. The Zero Discharge of Hazardous Chemicals, ZDHC, industry initiative, was activated by brands in response to this campaign [44]. More recently, several campaigns and publications to fight microplastics and plastic pollution in the ocean have acquired significant attention.

Policy Development

In some cases, countries, states or municipalities have developed own legislation or regulations to deal with transparency issues in the private sector. These governmental mandates enable the local industry as well as investors to set their basic standards on a much more levelled playing field [5], ensuring compliance to legal minimum requirements is accomplished. Certainly, regulatory requirements have to be paired up with effective control over compliance with the law to ensure that business behaviour will change. Some examples of policy and legislation development are the California Transparency in Supply Chains Act of 2010; the UK Modern Slavery Act of 2015; the French Duty of Vigilance Law of 2017; the National Pact for the Eradication of Slave Labour voluntary initiative in Brazil of 2005; and the Due Diligence on Child Labour Act in The Netherlands [9, 47]. Some of these regulations are also in accordance with International Organisation guidelines such as the OECD Due Diligence Guidance for Responsible Business Conduct, the OECD Due Diligence Guidance for Responsible Supply Chains in the Garment and Footwear Sector and the ILO International Labour Standards. All of the above should ultimately encourage better labour and environmental standards striving towards more sustainable value chain practices [45].

Corporate Initiatives

As stated above, many corporate practices are initiated as a response to public campaigns which spotlight poor labour, social or environmental standards throughout a company's value chain. These responses usually take the form of voluntary codes of conduct or supplier information disclosure [5]. To the date, several companies, such as Nudie Jeans, Patagonia and Icebreaker, have access on their websites to supplier maps, presenting an overview of the manufacturing facilities they work with, and the sourcing of their raw material [14, 39, 43]. This information helps build trust between stakeholders as well as to enhance brand reputation through marketing [33]. It also allows human rights and environmentalist organisations to take a closer look at different companies' value chains in order to assess their social and environmental performance. Nevertheless, the need for comparable and consistent information throughout the whole industry still remains one of the main challenges of transparency practices, as today's disclosure practices depend on corporate decision-making which differs from company to company [9].

Another interesting type of corporate initiatives to point out is collaborative benchmarking tools. An example around the issues of human rights is the Corporate Human Rights Benchmark, CHRB, developed in cooperation between the private sector, human rights organisations and investors [10]. It mainly consists of a public scorecard for T&C, agriculture and extractive value chain practices on human rights issues (ibid.). The extensive use of self-assessment tools such as the Sustainable Apparel Coalition's, SAC, Higg Index has made more relevant for brands to start tracing their products throughout their value chain. This tool enables brands to assess their social, labour and environmental performance against industry standards [2].

5.3 Main Implementation Challenges

There are three main challenges identified throughout this chapter regarding the implementation of transparency practices. Firstly, information disclosed or made visible is usually inconsistent or incomparable between different businesses and organisations. There is a need to set minimum disclosure standards to pursue a common understanding of transparency practices regarding complete supplier information, scope of the disclosure, frequency of information updates and formats for disclosure [9]. In this regard, corporate initiatives such as benchmarking tools may aid the achievement of a common language to make information released through transparency practices comparable and consistent between each other. Secondly, still many businesses are taking non-disclosure decisions based on reasons of loss of competitive advantage and sensitive information sharing for the business [19]. On these grounds, many pre-competitive industry collaborations or anti-competition regulations may motivate more companies into a more transparent disclosure of relevant information. Thirdly, lack of transparency between value chain partners also leads to information inefficiency challenges. The most common outcomes of this implementation challenge are discrepancies between supply and demand, which become more and more distorted when moving further away from the market [38]. This challenge grows exponentially due to power and involvement differences and struggles between partners. In order to tackle this, ICT networks, linked to consumer demands, which enable information sharing in a simple manner may help reduce the discrepancies within the network (ibid.). These authors state, however, that the implementation of these networks will be greatly dependent upon a willingness to share information and the partners' access to technological processes and products.

6 A Way Forward

The industry, academia and supporting organisations currently agree that traceability and transparency are prerequisites to shift to sustainable consumption and production patterns, as defined through SDG 12. However, more massive uptake of these practices is needed to accelerate this shift. The chapter presents an overview of current practices and industry examples which relate SDG 12 targets to transparency and traceability practices. It expects to provide a clear understanding of the concepts as well as to motivate the uptake of these practices.

References

1. Agrawal T, Koehl L, Campagne C (2018) A secured tag for implementation of traceability in textile and clothing supply chain. Int J Adv Manuf Technol 99(9–12):2563–2577
2. Apparelcoalition.org (2019) The Higg Index—Sustainable Apparel Coalition. [online] Available at: https://apparelcoalition.org/the-higg-index/
3. ASKET (2019) Full traceability—garment tracing, from farm to finish line. [online] Available at: https://www.asket.com/traceability/
4. Bhaduri G, Ha-Brookshire J (2011) Do transparent business practices pay? Exploration of transparency and consumer purchase intention. Cloth Textiles Res J 29(2):135–149
5. Birkey R, Guidry R, Islam M, Patten D (2016) Mandated social disclosure: an analysis of the response to the California Transparency in Supply Chains Act of 2010. J Bus Ethics 152(3):827–841
6. Brundtland G (1987) Our common future: report of the 1987 World Commission on Environment and Development. United Nations, Oslo, pp 1–59
7. Carter CR, Rogers DS (2008) A framework of sustainable supply chain management: moving toward new theory. Int J Phys Distrib Logist Manage 38:360–387
8. Circle Economy (2019) The circularity gap report 2019. The platform for accelerating the circular economy (PACE). [online] Available at: https://www.circularity-gap.world/. CC by 1.0
9. Clean Clothes Campaign (2017) Follow the thread: the need for supply chain transparency in the garment and footwear industry
10. Corporate Human Rights Benchmark (2019) Home. [online] Available at: https://www.corporatebenchmark.org/
11. Echa.europa.eu (2019) Understanding REACH - ECHA. [online] Available at: https://echa.europa.eu/regulations/reach/understanding-reach
12. Egels-Zandén N, Hulthén K, Wulff G (2014) Trade-offs in SC transparency: the case of Nudie Jeans Co.
13. ElMessiry M, ElMessiry A (2018) Blockchain framework for textile supply chain management. Lect Notes Comput Sci 213–227
14. eu.icebreaker.com (2019) Icebreaker—merino wool clothing for outdoor and performance sports. [online] Available at: https://eu.icebreaker.com/en/transparency.html
15. Fashion Revolution (2019) Home—Fashion Revolution. [online] Available at: https://www.fashionrevolution.org/
16. Fashionunited.nl (2019) [online] Available at: https://fashionunited.nl/images/201804/4AMadeinlabel1.jpg
17. Fletcher K (2008) Sustainable fashion and textiles: design journeys
18. Flickr (2019) Sunday telegraph Australia day hat—made in China—one size fits all. [online] Available at: https://www.flickr.com/photos/angusf/2270317189

19. Global Fashion Agenda (2018) CEO agenda 2018: seven sustainability priorities for fashion industry leaders
20. Gray R, Javad M, Power DM, Sinclair CD (2001) Social and environmental disclosure and corporate characteristics: a research note and extension. J Bus Financ Acc 28(3–4):327–356
21. Greenpeace International (2019) Detox My Fashion—Greenpeace International. [online] Available at: https://www.greenpeace.org/international/act/detox/
22. Gs1.org (2019) Annual report 2018. [online] Available at: https://www.gs1.org/sites/default/files/docs/annual_report/GS1-Annual-Report-2018.pdf
23. Gunasekaran A, Hong P, Fujimoto T (2014) Building supply chain system capabilities in the age of global complexity: emerging theories and practices. Int J Prod Econ 147:189–197
24. Hu J, Zhang X, Moga LM, Neculita M (2013) Modeling and implementation of the vegetable supply chain traceability system. Food Control 30:341–353
25. Iso.org (2018) Traceability [online] Available at: https://www.iso.org/obp/ui/iso:std:iso:9000:ed-4:v1:en:term:3.6.13
26. Joy A, Peña C (2017) Sustainability and the fashion industry: conceptualizing nature and traceability. In Sustainability in fashion, pp 31–54
27. Kamal Y, Deegan C (2013) Corporate social and environment-related governance disclosure practices in the textile and garment industry: evidence from a developing country. Aust Acc Rev 23(2):117–134
28. Kang J, Hustvedt G (2014) The contribution of perceived labor transparency and perceived corporate giving to brand equity in the footwear industry. Cloth Textiles Res J 32(4):296–311
29. Kumar V, Koehl L, Zeng X (2016) A fully yarn integrated tag for tracking the international textile supply chain. J Manuf Syst 40:76–86
30. Kumar V, Hallqvist C, Ekwall D (2017a) Developing a framework for traceability implementation in the textile supply chain. Systems 5(2):33
31. Kumar V, Agrawal T, Wang L, Chen Y (2017b) Contribution of traceability towards attaining sustainability in the textile sector. Textiles Cloth Sustain 3(1)
32. Linich D (2014) The path to supply chain transparency. A practical guide to defining, understanding, and building supply chain transparency in a global economy
33. Ma Y, Lee H, Goerlitz K (2015) Transparency of Global Apparel supply chains: quantitative analysis of corporate disclosures. Corp Soc Responsib Environ Manag 23(5):308–318
34. Macchion L, Furlan A, Vinelli A (2017) The implementation of traceability in Fashion Networks. In Collaboration in a data-rich world, pp 86–96
35. Madaster.com (2019) Madaster origination: Madaster. [online] Available at: https://www.madaster.com/en/about-us/why-a-materials-passport
36. Mani V, Gunasekaran A, Delgado C (2018) Supply chain social sustainability: standard adoption practices in Portuguese manufacturing firms. Int J Prod Econ 198:149–164
37. Martin DM, Schouten J (2011) Sustainable marketing. Pearson Prentice Hall, New York
38. Minami C, Nishioka K, Dawson J (2012) Information transparency in SME network relationships: evidence from a Japanese hosiery firm. Int J Logist Res Appl 15(6):405–423. https://doi.org/10.1080/13675567.2012.749848
39. Nudiejeans.com (2019) Production guide—Nudie Jeans. [online] Available at: https://www.nudiejeans.com/productionguide/
40. OECD (2013) Trade policy implications of global value chains. http://www.oecd.org/sti/ind/Trade_Policy_Implications_May_2013.pdf
41. OECD (2014) Greening household behavior: overview from the 2011 survey, revised edn. OECD Studies on Environmental Policy and Household Behaviour
42. Panigrahi SS, Rao NS (2018) A stakeholders' perspective on barriers to adopt sustainable practices in MSME supply chain. Issues and challenges in the textile sector
43. Patagonia.com (2019) The Footprint Chronicles. [online] Available at: https://www.patagonia.com/footprint.html
44. Roadmaptozero.com (2019) ZDHC. [online] Available at: https://www.roadmaptozero.com/
45. Seuring S, Müller M (2008) From a literature review to a conceptual framework for sustainable supply chain management. J Clean Prod 16(15):1699–1710

46. Shea A, Nakayama M, Heymann J (2010) Improving labour standards in clothing factories. Glob Soc Policy Interdisc J Publ Policy Soc Dev 10(1):85–110
47. Stevenson M, Cole R (2018) Modern slavery in supply chains: a secondary data analysis of detection, remediation and disclosure. Suppl Chain Manage [Online] 23(2): 81–99
48. Sustainabledevelopment.un.org (2019a) Goal 12: sustainable development knowledge platform. [online] Available at: https://sustainabledevelopment.un.org/sdg12
49. Sustainabledevelopment.un.org (2019b) Transforming our world: the 2030 agenda for sustainable development: sustainable development knowledge platform. [online] Available at: https://sustainabledevelopment.un.org/post2015/transformingourworld
50. Svensson G (2009) The transparency of SCM ethics: conceptual framework and empirical illustrations. Supply Chain Manag Int J 14(4):259–269
51. Textileexchange.org (2019) Glossary and abbreviations | Textile Exchange. [online] Available at: https://textileexchange.org/materials/glossary/
52. TrusTrace.com (2019) TrusTrace | trust through traceability. [online] Available at: https://www.trustrace.com/
53. Unece.org (2019) Trade—UNECE. [online] Available at: https://www.unece.org/tradewelcome/outreach-and-support-for-trade-facilitation/traceability-for-sustainable-value-chains-textile-and-leather-sector.html
54. United Nations (2011) Guiding principles on business and human rights: implementing the United Nations 'Protect, Respect and Remedy' framework. HR/PUB/11/04
55. United Nations (2016) Traceability for sustainable trade: a framework to design traceability systems for cross border trade. ECE/TRADE/429. Issued by the Economic Commission for Europe, UN/CEFACT
56. United Nations (2017) TEXTILE4SDG12: transparency in textile value chains in relation to the environmental, social and human health impacts of parts, components and production processes. ECE/TRADE/439. Issued by the Economic Commission for Europe, UN/CEFACT
57. Vishwanath T, Kaufmann D (2001) Toward transparency: new approaches and their application to financial markets. World Bank Research Observer 16:41–57
58. Webster K (2016) The circular economy: a wealth of flows. Ellen MacArthur Foundation Publishing, Isle of Wight
59. World Bank (2017) ICT in agriculture: connecting smallholders to knowledge, networks, and institutions, updated edn. World Bank, Washington, DC. https://doi.org/10.1596/978-1-4648-1002-2. License: Creative Commons Attribution CC BY 3.0 IGO

Sustainable Development Goal 12 and Its Relationship with the Textile Industry

Marisa Gabriel and María Lourdes Delgado Luque

Abstract The United Nations Sustainable Development Goals set the course: the 2030 Agenda for Sustainable Development. Out of these 17 goals—also known as SDG—number 12 refers to sustainable production and consumption. It basically means a reduced use of resources and/or their correct management. The circular economy, for its part, aims to continuously keep products, components, and materials at their highest value. It proposes a long-term system in which today's goods can become tomorrow's materials, availing of prudence and equity to reconcile development and economy with environment and society. In this way, industrial processes are no longer a threat to the ecosystem, but, on the contrary, they seek to revalorise resources, thus promoting sustainable development. The textile industry is an essential part of people's everyday life and a very important sector in the global economy. Therefore, the purpose of this chapter is to analyse SDG 12 and how it can be applied to the textile industry, considering the circular economy as a way towards sustainable development. In this connection, the chapter begins with an introduction to the SDGs, particularly to SDG 12, continues presenting the concepts of circular economy and textile industry, and concludes with the association of those concepts with cases such as Dutchawearness (the Netherlands), Excess Materials Exchange (the Netherlands), Rapanui (UK), Stylelend (New York), and Tejidos Royo (Spain). The chapter draws a few conclusions at the end, which could be summarised by stating that, to bring about a change of outlook, the circular economy should be at the heart of the company, reorganising resources, assets, capital and, above all, business potential to ensure the future of the company, which is the only way to achieve SDG 12.

Keywords Sustainable development goals · Textile industry · Circular economy · Design

M. Gabriel (✉)
Sustainable Textile Center, Paroissien 2680, 5th "B", C1429CXP Buenos Aires, Argentina
e-mail: maragabriel6@gmail.com

M. L. D. Luque
Sustainable Textile Center, Imperio Argentina 12, gate 2, 4th B, 29004 Malaga, Spain

M. A. Gardetti and S. S. Muthu (eds.), *The UN Sustainable Development Goals for the Textile and Fashion Industry*, Textile Science and Clothing Technology,
https://doi.org/10.1007/978-981-13-8787-6_2

1 The United Nations Sustainable Development Goals

1.1 Background: The Millennium Declaration (Millennium Development Goals), the Global Compact, and the Principles for Responsible Management Education (PRME)

At the turn of the century, the United Nations achieved global commitment with the "Millennium Declaration" [22]. The Millennium Declaration signed by 189 Heads of State in year 2000 translated into an opportunity for developed and developing countries to undertake the commitment to be compliant with eight development goals—called the "Millennium Development Goals"—that had to be attained by 2015. According to Fuertes [6], the Millennium Declaration meant the possibility of changing the approach of public policies, in that the economic system had to be subordinate to broader social goals focused on human development as the basic core of development.

Arising from the "Millennium Declaration", the Global Compact is a joint initiative of the United Nations Development Programme (UNDP), the Economic Commission for Latin America and the Caribbean (ECLAC), and the World Labour Organization (WLO). By means of business voluntary commitment, the Global Compact promotes a new corporate culture on how to manage businesses.

The main purpose of the Global Compact is to enable the development of corporate social responsibility, fostering human rights, labour standards, environmental protection, and anti-corruption.

Its real essence is to create an ever-growing labour network supporting businesses through learning and knowledge sharing, exercising leadership as a corporate citizen, and hence exerting influence on others through their behaviour [5]. In a few words, the Global Compact is the contribution of the private sector to the Millennium Goals [11] and, therefore, to the Sustainable Development Goals or the 2030 Agenda. The goal of the Global Compact is to help align corporate policies and practices to universally concurred and internationally applicable ethical goals, which are inspired in the Universal Declaration of Human Rights, Declarations of the International Labour Organisation on Fundamental Principles and Labour Rights, and the Declaration of Rio of the United Nations Conference on Environment and Development.

At the Leaders' Summit 2007, the United Nations Global Compact Office presented the Principles for Responsible Management Education—PRME—at business schools and academic institutions. Their purpose is to improve the training of future business leaders in social issues, human rights, and environmental protection. Said principles have been concurred by several academics from business schools and academic associations around the world, and they are intended to create a framework laying the foundations for common and integrated education within a society that is becoming more and more globalised and which needs new values for a more sustainable development of the world.

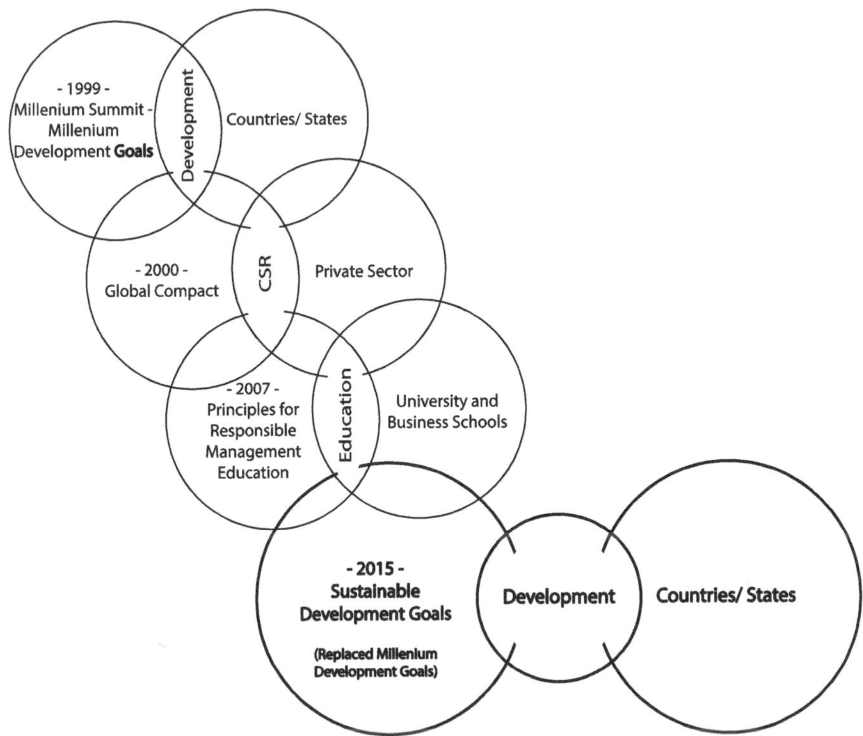

Fig. 1 Path leading to the sustainable development goals. *Source* Prepared by the authors

Figure 1 shows the path leading to the Sustainable Development Goals. On the left side, it shows the different initiatives and, on the right side, the sectors involved.

1.2 The Sustainable Development Goals or the 2030 Agenda

In 2012, Rio de Janeiro Summit, organised by the United Nations and called "Rio+20," analysed the progress made since the summit held in Rio de Janeiro in 1992, and it also announced that the Millennium Goals were to be replaced—starting in 2015—by the Sustainable Development Goals, also called "the 2030 Agenda for Sustainable Development Goals".

In September 2015, the United Nations General Assembly approved the agenda, which sets up a transformational view to economic, social, and environmental sustainability. For the 193 signatory Member States, it is—now and until 2030—the reference guideline for the institution's work towards this vision. The reason for this is that the slow global growth, social inequalities, and environmental depletion—typical of our current reality—pose unparalleled challenges to the international community. In fact, we are facing the change of an era: as the alternative of continuing with the same patterns is no longer viable, the current development paradigm should be transformed into an inclusive paradigm based on sustainable development and with a long-term vision. The Agenda comprises the 17 goals below, which, in turn, include 169 targets[1]:

Goal 1: End POVERTY. in all its forms everywhere.
Goal 2: Zero HUNGER. End hunger, achieve food security and improved nutrition and promote sustainable agriculture.
Goal 3: Good HEALTH. Ensure healthy lives and promote well-being for all at all ages.
Goal 4: Quality EDUCATION. Ensure inclusive and equitable quality education and promote lifelong learning opportunities for all.
Goal 5: Gender EQUALITY. Achieve gender equality and empower all women and girls.
Goal 6: Clean WATER and sanitation. Ensure availability and sustainable management of water and sanitation for all.
Goal 7: Affordable and clean ENERGY. Ensure access to affordable, reliable, sustainable and modern energy for all.
Goal 8: Decent WORK and economic growth. Promote sustained, inclusive and sustainable economic growth, full and productive employment and decent work for all.
Goal 9: INDUSTRY, innovation, infrastructure. Build resilient infrastructure, promote inclusive and sustainable industrialisation and foster innovation.
Goal 10: Reduced INEQUALITIES. Reduce inequality within and among countries.
Goal 11: SUSTAINABLE CITIES AND COMMUNITIES. Make cities and human settlements inclusive, safe, resilient and sustainable.
Goal 12: Responsible CONSUMPTION and production. Ensure sustainable consumption and production patterns.
Goal 13: CLIMATE Action. Take urgent action to combat climate change and its impacts.
Goal 14: Life BELOW WATER. Conserve and sustainably use the oceans, seas, and marine resources for sustainable development.
Goal 15: Life ON LAND. Protect, restore, and promote sustainable use of terrestrial ecosystems, sustainably manage forests, combat desertification, and halt and reverse land degradation and halt biodiversity loss.
Goal 16: Peace, JUSTICE and strong institutions. Promote peaceful and inclusive societies for sustainable development, provide access to justice for all and build effective, accountable and inclusive institutions at all levels.
Goal 17: PARTNERSHIP for the goals. Strengthen the means of implementation and revitalise the global partnership for sustainable development.

[1]For more information, please visit: http://www.undp.org/content/undp/es/home/sustainable-development-goals.html.

1.3 Sustainable Development Goal 12: Summary

This goal—which is the purpose of this chapter—refers to sustainable production and consumption. These refer to the efficient use of natural resources and energy, development of environmentally friendly infrastructure, improved access to basic services, and creation of fair paid jobs under good labour conditions. This helps improve quality of life and devise development plans to reduce economic costs and both environmental and social impacts.

As sustainability is systemic, a broad approach should be used to analyse the entire corporate supply chain, from product to final disposal by consumers, including, in turn, consumer education by providing knowledge about responsible consumption and sustainable ways of living.[2]

Sustainable Development Goal 12 basically means a reduced use of resources, but—above all—efficient resource management, moving away from old production, and consumption patterns. Everything that we produce and consume leaves a footprint; therefore, this particular goal is focused on reducing such footprint. This can be achieved by reconsidering the operation of the economy based on an analysis of product life cycle and product life cycle/market relationship. The circular economy is an alternative to achieve this. Therefore, Lacy and Rutqvist [15] refer to the circular economy as a strategy to reorganise the industry and take a fresh approach to both the market and the product/consumer relationship. It is precisely about turning waste into strengths and/or wealth, with a particular emphasis on the need to reconsider the squandering of natural resources, products, and goods.

2 The Circular Economy

2.1 Linearity and Circularity

"*The transition to a circular economy may be the biggest revolution and opportunity for how we organize production and consumption in our global economy in 250 years*" [15: xv]. For such purpose, we should take a radically fresh approach to the relationship among market, producers, consumers, and natural resources in order to consume and produce in a disruptive way based on innovative business models conducive to a fairer industry.

The economic, production, and consumption system brought about by the industrial revolution—and still in place—is regarded as a "linear system". This system is based on the extraction of raw materials, which are modified to get an end product that is sold on the market and purchased by consumers who will later either dispose of them or give them another use. This process is illustrated in Figure 2.

[2]For more information, please visit: https://www.globalgoals.org/12-responsible-consumption-and-production.

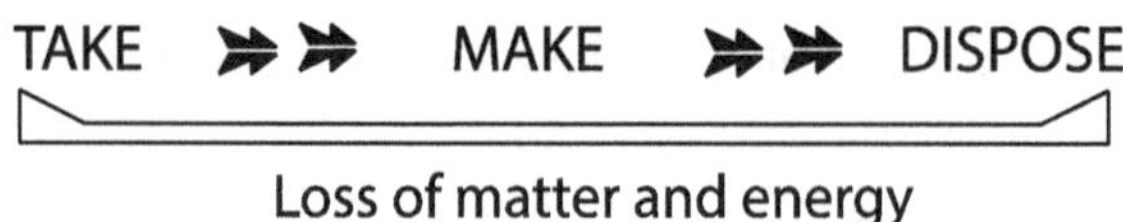

Fig. 2 Linear economic system. *Source* Prepared by the authors

This linear process, ruled by consumption, was taken to its ultimate expression by creating goods that last increasingly less through planned obsolescence,[3] which, in turn, creates fictitious needs in consumers.

Current concerns over climate change, population growth, and resource scarcity evidence the dangers of continuing with the linear economy. Therefore, the circular economy emerges as an alternative, as a way to holistically consider production and consumption based on a regenerative proposal.[4] The circular economy promotes the efficient use of resources, thus reducing raw material extraction and maximising its useful life while increasing production, based on innovative ideas and enabling technologies. In other words, it is an economic system in which both matter and energy circulate, reducing the influence of human activity on the environment. Matter flows through closed circuits and gets transformed, making an efficient use of the necessary energy [17]. And here, the role of designers is key. You should be innovative and capable of thinking of out the box of our current knowledge and product creation structure. While linearity proposes simple, mechanical, efficient and already proven systems, a circular system depends on the efficiency of complex, adjustable, and interdependent systems; however, it requires, above all, commitment, openness, and

[3]Planned obsolescence shortens product life in order to encourage consumers to replace products; thus, consumption becomes an industry driver and creating more jobs. When the concept was first coined, it was based on the wrong idea that natural resources were infinite.

[4]The background to the circular economy includes various schools and lines of thought, which evolved into different circular economy models. These currents are characterised by their focus on nature and emerge as urgent calls for a change of paradigm to offer sustainable development models. Among them, we can mention:

- "Cradle to Cradle" [18] proposed to extend the useful life of materials while sorting components—previously regarded as waste—into biological or technical components.
- Regenerative Design, which means to regenerate energy and matter during the production process.
- Industrial Ecology, which, according to Graedel and Allenby [12], is "*the means by which humanity can deliberately and rationally approach and maintain a desirable carrying capacity, given continued economic, cultural, and technological evolution. The concept requires that an industrial system be viewed not in isolation from its surrounding systems, but in concert with them. It is a systems view in which one seeks to optimize the total materials cycle from virgin material, to finished material, to component, to product, to obsolete product, and to ultimate disposal. Factors to be optimised include resources, energy, and capital*" [12: 9].
- Blue Economy, which is based on physics, using natural systems that cascade nutrients, energy, and materials. Gravity is the main source of energy, solar energy is the natural fuel, and water is the primary solvent.
- We could also mention biomimicry [2] and permaculture [19].

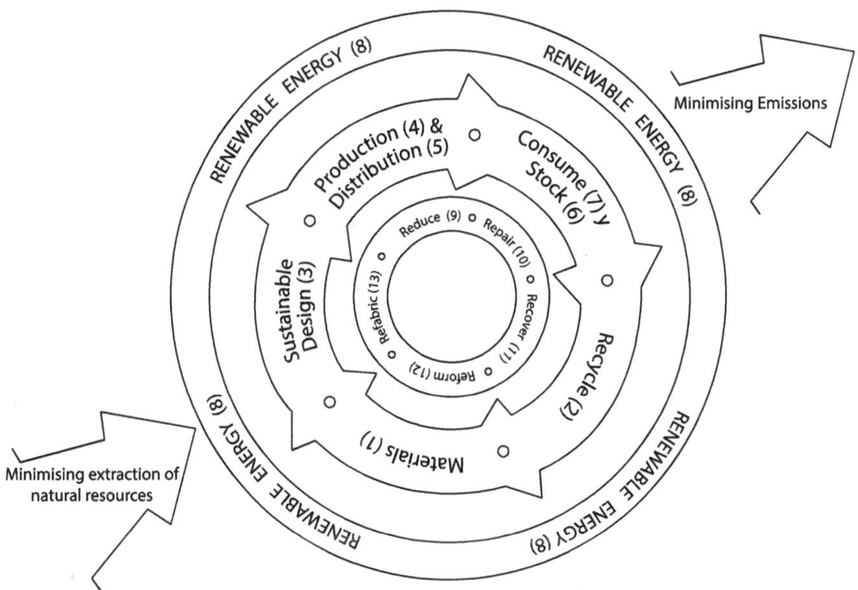

Fig. 3 Circular economy model. *Source* Prepared by the authors

the ability to take responsibility for the impact of individual actions and decisions on nature and the environment.

It is not just about replacing a product with a less harmful one, changing a raw material or the type of energy used for processing, but also about thinking whether the corporate business model is sustainable or not.

Catherine Weetman, in her book A Circular Economy Handbook for Business and Supply Chain [30], states that circular economy is the ideal concept to define a truly sustainable economy that works without waste, saves resources, and is in synergy with nature. Along this line, Ray Anderson [1], founder of Interface, argues that nature does not waste; one organism's waste is—and should be—another's food.

Figure 3 shows a circular system scheme that offers a simple view of the links that make up the circular production/consumption chain. On the one hand, we should minimise natural resource extraction, using materials already available on the market both as resources and as raw materials. In this connection, we should rethink materials (see No. 1, in Fig. 3) and recycling (see No. 2, in Fig. 3), how we disassemble products and recover the different post-consumer materials and, in some cases, we should also replace raw materials, taking a fresh approach to product design and production from the start.

This figure also mentions sustainable design (3), when referring to such design planned and carried out in a disruptive way, prioritising respect for the environment and the society. On the other hand, responsible production (4), distribution (5), stock management (6) and consumption (7) practices should be implemented, avoiding the overuse of energy, and the manufacture of more products than those demanded by the market. In this connection, while consumers are offered longer life products, they should be educated to refrain from replacing a product with a newer one quickly. The use of renewable energy (8) sources for processing should always be intended. To implement a circular economy system, it is essential to avoid the use of fossil fuels or non-renewable energies. To achieve the above, the centre of the figure shows product reduce (9), repair (10), recover (11), reform (12), and remanufacture (13). In the meantime, the industry would minimise gas emissions. As a result, the circular economy emerges as a way to achieve SDG 12.

While transitioning and migrating models, small successes are vital, since businesses should gradually build the migration path towards a circular economy.

According to Webster [29], different rules are required if human beings still want to feel "at home" in the modern world. The author proposes that the circular economy is an expression of intelligent systems, an opportunity for a broader and widespread update of both the economy and businesses, where—more importantly—the system is enriched. It is about a transformation that should begin at the core of the system and businesses.

2.2 *Definition of Circular Economy*

The circular economy aims to continuously keep products, components, and materials at their highest value. It is a long-term system in which today's goods are tomorrow's materials.

The circular economy avails of prudence and equity to reconcile development and economy with environment and society. Industrial processes are no longer a threat to the ecosystem, but, on the contrary, they seek to revalorise resources, thus promoting sustainable development. It is a new way to look at the relationship between market/consumers and natural resources, directly focused on Sustainable Development Goal 12.

Tonelli and Cristoni [27] argue that it means to keep a continuous flow of products, components, materials, and energy, minimising their environmental footprint. Products should either re-enter the production cycle or biodegrade.

This concept should be at the heart of the company, which should focus on "*devising regenerative strategies and actions for creating a closed-loop production and consumption system*" [27: 39].

To this end, the product life cycle should be analysed. This requires design, innovation, science, technology, cooperation, and—above all—education. "*The voyage of discovery lies not in seeking new horizons, but in seeing with new eyes*" [29: 18].

2.3 The Pillars of the Circular Economy

According to Webster [29], the circular economy is based on five principles:

- Design out waste. This means taking waste as raw material, and thinking how it could be dismantled and reused so that it is not discarded again. With this approach, waste is turned out into biological (biodegradable) or technical (reusable) material. Products are usually divided into a biological trace component and a reusable material component.
- Build resilience through diversity. We should stop looking for efficiency in the current models and think out of the box, working in process adaptability, modularity, and versatility based on diverse, interconnected systems.
- Use renewable energy. Minimise the use of fossil fuels and ensure the efficient use of energy.
- Think in systems. The key to apply or implement the circular economy is to think in systems. Understanding the influence and interconnection of the parts and the whole is critical.
- Think in cascades. This means to get the highest value out of products and materials in every step of the process.

For Tonelli and Cristoni [27], the circular economy is based on four principles, namely:

- Embrace green technologies and focus on the responsible use of natural inputs to produce.
- Maximise utilisation rate of company assets. It is about maximising assets and product life once it is in the market, by fully exploiting and maximising utilisation rates through innovative solutions and turning waste into raw materials for other businesses.
- Circulate goods, product components, and materials based on recycling, reuse, and remanufacturing in order to keep the value at their highest.
- Minimise and gradually phase out negative externalities, that is, environmental and social damages (If the first three principles are met, externalities would be reduced, leading to zero negative impact).

Tonelli and Cristoni's principles [27] could be translated into business objectives so that circularity is easier to both analyse and achieve. This could mean to regenerate, share, optimise, loop, virtualise, and exchange. And, in turn, those objectives could be applied to every business area—reverse cycle innovation and design; operation and management of technologies to improve ecological practices; development of provider trust and commitment; internal alignment and external cooperation—, adjusting and developing new circular economy-enabling technologies.

In short, from the business perspective, this is about thinking of products not just as an end in themselves, but as an opportunity for further value creation and long-term customer relationships. Businesses should make consumers embrace circular consumption patterns, since the circular system operation also depends on them.[5] From a consumption perspective, as it is systemic, the system is no longer exclusively dependent on the act of purchase, but it needs the users and their feedback. This is the reason why businesses should educate consumers.

Charter [3] points out that the circular economy not only offers long-term benefits for the business, but it also maximises capabilities and recovers resources, keeping the value at its highest. Businesses reduce both waste and the use of resources. To operate with a circular economy system, it is vital to count on the support from every player in the economy, along with government policies.

3 The Textile Industry

The textile industry is a vital part of people's everyday life and a very important sector in the global economy. It is an essential industry, since we could not imagine a society without apparel. However, while it could make a significant contribution and promote sustainability, it is positioned as one of the most polluting industries.

With a few exceptions, this industry operates within a linear, obsolete system. It is an extraction, production, and consumption system with large cracks that reveal and highlight social and environmental issues. Far from repairing damages, this industry continues with its legacy mandate, even embracing increasingly less sustainable models. "*The textile industry uses large quantities of water and energy (two of the most pressing issues worldwide), in addition to building up waste, effluents and pollution. Both textile product manufacture and consumption—whether in fashion or not—are significant sources of environmental damage*" [9: 110].

In the textile and fashion universe, fast fashion businesses took the linear production and consumption system to its ultimate expression. The past two decades were crucial. The rapid growth and expansion of these businesses would take the industry to a truly unsustainable situation. "*In the last 15 years, clothing has approximately doubled, driven by a growing middle-class population across the globe and increased per capita sales in mature economies. The latter rise is mainly due to the "fast fashion" phenomenon, with quicker turnaround of new styles, increased number of collections offered per year, and—often—lower prices*" [20: 18].

In some cases, this translates into the release of up to 52 collections a year, with no regard to environmental or social issues. However, authors argue that we are on the verge of a new age—the age of fast fashion sustainable brands, since "*fast fashion brands are beginning to show a shift towards changes in their manufacturing processes*" [26: 22].

[5] Businesses should inspire, educate and create programmes to encourage consumers to embrace more sustainable practices. For example, at product end of life, they should make consumers return products to the store or fix them if damaged.

If even the most socially and environmentally committed businesses are believed to be currently looking for a change towards a more sustainable operation, it is time to fully rethink the industry.

3.1 The Textile Industry Problems

The textile industry generates waste and effluents in every stage. The most basic relation that could be established between waste and the textile industry would be garment cutting trimmings or post-consumer disposed garments; however, pollution is present in every link of the production chain. Waste can be classified into three subtypes: pre-consumption waste (trimmings and raw material remnants), post-consumption waste (post-use disposed garments), and post-industrial waste, i.e., related to dyes, finishing processes, chemical waste, and environmental pollution caused, for example, by transport [10, 25, 28].

The first link of the production chain—sourcing of natural or synthetic textile fibres—has an impact on the end product environmental footprint, which is more detrimental in some cases than in others. Throughout this process, natural fibres are swollen and washed with large amounts of water and, sometimes, chemical solvents, which pollute soil, water and rivers, and impair the surrounding communities. Moreover, synthetic fibres, such as petrol-derived fibres, have an impact too.

> Considering the whole textile chain –from spinning to consumer use-, it cannot be overlooked that the use of chemicals may have carcinogenic and neurological effects, may cause allergies and may affect fertility. During these two processes, large amounts of water and energy are used and, in general, non-biodegradable wastes are produced. [8: 106]

Global Fashion Agenda and Boston Consulting Group analysed the damages caused by this industry, the forecast, and future prospects [16]. They submitted a report with the damages currently caused by the textile industry, and how they will scale up by 2050. This assessment analysed those problems, and businesses were called on to become active change agents; otherwise, forecasts show irreversible damages and potential industry shutdown due to resource scarcity and environmental pollution:

- In the textile industry, water is a key and ever-present resource. This resource is required from fibre growing to processing and garment wear. Based on the current practices, annual water use was estimated to increase in 50% by 2030.
- Carbon dioxide emissions, especially during processing stages, are even a worse problem. Atmospheric CO_2 is 20% over the safe level, and industrial emissions are expected to increase in 60% by 2030.
- Chemicals are more difficult to track and evaluate. This industry uses fertilisers, pesticides, colourants, pigments, dyes, or even processing agents. The analysis of occupational diseases caused by carcinogens and airborne particles makes evident that this industry requires advanced chemical management.

- Waste is far more than a problem in this industry. If the current practices continue, by 2030, the annual production and consumer disposal waste will increase in 60%. Only 20% of post-consumer clothes is recycled, and most of these garments lose value due to inappropriate processing technologies and ignorance. In this sense, we believe that the industry will have a huge advantage if it starts working with discarded materials.

3.2 Textile and Clothing Production Chain

Along this line, it is vital to rethink the entire cycle if we want to present and ensure a sustainable closed-loop consumption and production system for the textile industry making connections, building bridges and bonds, and creating a systemic and collaborative system. This industry is featured by long production and supply chains. Therefore, we will try to describe every step of the textile industry so as to understand how to create a circular system suitable for this industry.

Figure 4 shows the production chain links from raw material sourcing to post-use disposal. It should be noted that, most times, the steps of these processes take place in different countries; hence, significant amounts of fuel and energy are used for transport.

The first step is to get the initial raw materials, i.e. textile fibres, whether natural or synthetic. Fabrics can be manufactured after the extraction of threads that will be used to make the yarns. Yarns are woven into meshes and, finally, they are subject to the dry-cleaning and finishing processes to add any desired property, which will be translated into the end product. At this point, we only have the fabric, which is used to make different products. Moreover, clothing goes through a like process—design, cut, and tailoring—and, in this case, clothes are subject to various finishing processes, such as garment printing, pleating, or ironing. Once a garment is made, it is ready for sale. Then, consumers purchase and wear garments, being in charge of wear and care (washing, ironing, and putting away) processes, until those garments are thrown away, either through donation or disposal. Figure 5 shows the negative impact and resources required by all these processes.

Just like in small textile (yarn and fabric) and garment production chains, there are only a few cases in the entire production chain in which both products—garments—and supplies—textiles—are manufactured in the same country and even fewer cases with a proven transparent supply chain.[6] Notwithstanding that, different cases are coming to light. For example, P.I.C Style is a women's clothing brand that manufactures and uses raw materials and textiles made in the UK. This brand also offers a nine-item capsule wardrobe to create up to 50 looks and encourages responsible consumption.

In the above figures, design appears in two stages. Moreover, design should be considered and include the entire textile and clothing production process, as well as

[6] Source: https://pic-style.com/about-us/.

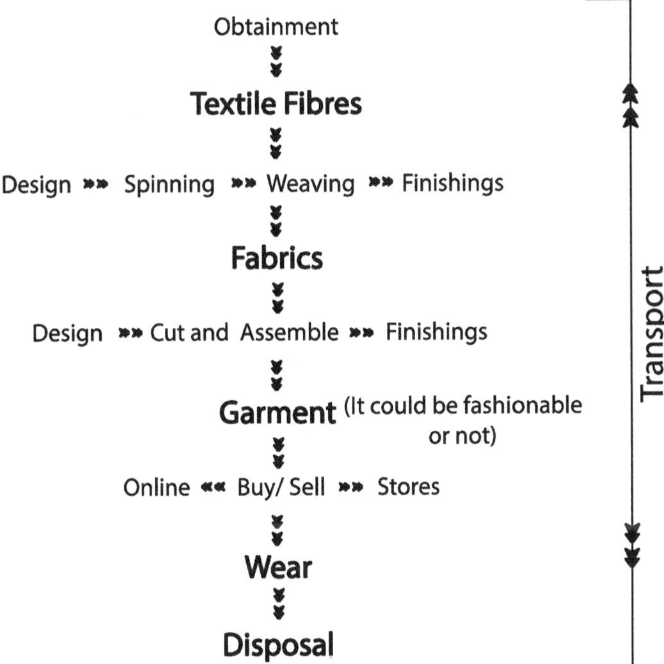

Fig. 4 Textile and clothing industry. *Source* Prepared by the authors

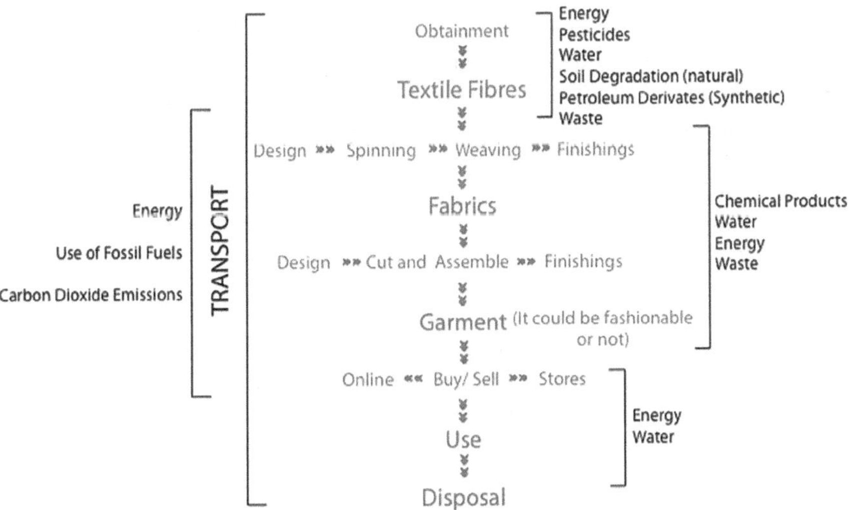

Fig. 5 Environmental Impacts of Textile and clothing industry. *Source* Prepared by the authors

wear and subsequent disposal. The reason for this is that design and innovation can help reshape this industry to reach all the target areas. In Victor Papanek's words [24: 4], design is, and should be, "*the conscious and intuitive effort to impose meaningful order*".

Design and, most specifically, design thinking will be definitely changed thanks to updated needs and possibilities and moving away from old models of thought. Moreover, all this calls for rethinking not only the product creation process, but also product life cycle.

3.3 The Circular Textile Industry Approach

Product life cycle begins even before fibre sourcing. Businesses should promote a circular economy and also, as stated by Galí [7: 343], "*They should also attribute a symbolic value to the object without reducing the product to its functionality*".

According to Niinimaki [23], we should design for longevity; businesses should actively engage consumers, creating new long-term relationships. The author analyses product life cycle, including production process, referred to as technical cycle, and considers that both cycles should be jointly closed and operated.

Figure 6 shows how to rethink a circular, regenerative, and collaborative model in the textile and clothing industry.

Ultimately, while consumers add symbolic value, businesses should be capable of committing to, encouraging, and inviting consumers to sustain such value. As a consequence, consumers should be educated, and brands should take care of their production models and inspire consumers as well, leading and guiding them along this change process—vital for a future compliant with the 2030 Agenda.

And all the above needs to be consistent with Zatonyi's thought [31: 27]: "*I wonder how many things we do not out of a supernatural metaphysical good, but out of a repeated imposition which may have had a reason before, but not now*".

In this connection, SDG 12 underscores society and consumer's roles. While garments, most specifically, fast fashion garments are items truly designed for quick disposal, obsolescence of fashion products—driven by aesthetic change and tied to changing social preferences—point out the psychological/social nature of those factors that affect fashion garment lifespans [4]. This is about thinking from beginning to beginning instead of from beginning to end. As stated by Hethorn and Ulasewicz [13: 129], "*It is time to approach production and economic processes from a different perspective, adjusting not just the ways of doing but also our ways of thinking*".

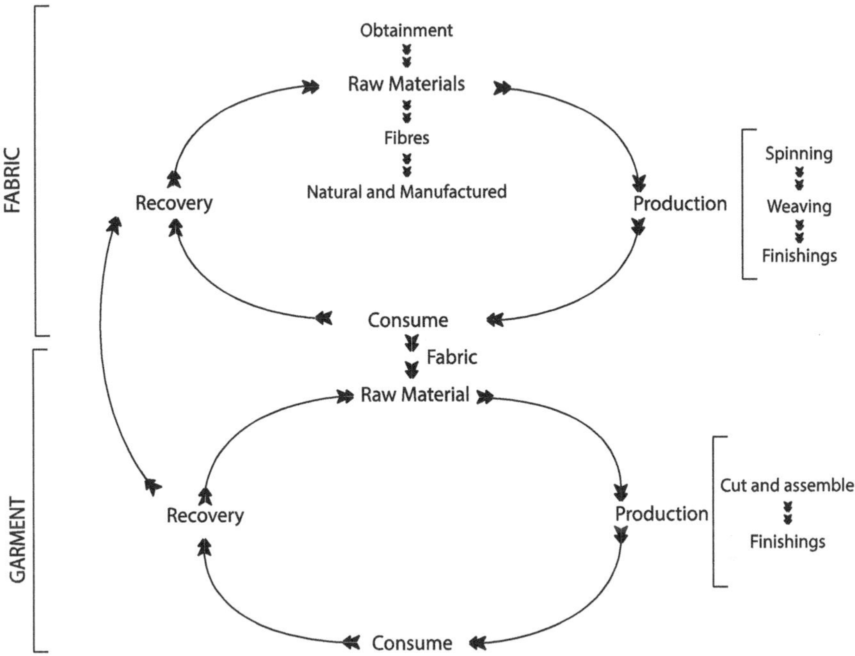

Fig. 6 Textile and clothing industry: circular economy model. *Source* Prepared by the authors

4 Application of the Circular Economy to the Textile Industry

For the successful implementation of the circular economy principles, businesses can choose from different models to meet their objectives. Tonelli and Cristoni [27] pose four strategies—which can be combined with one another—for businesses to apply the circular economy principles.

- Net Zero refers to a production process seeking to reduce the environmental impact. Here, the goal is to generate a neutral environmental footprint. In clothing, there are two ways to generate zero waste: one is to work and organise pattern-making based on total fabric width, and the other is to work using remnants from other products. For example, Tonlé is a brand that uses both ways. On the one hand, it sources scrap waste from clothing factories and, on the other, it cuts it into strips and ties them back together, working with patterns that do not generate waste. Then, it continues with clean production processes using zero chemicals and cuts the remaining fabric into tiny strips for weaving. The tiniest fabric and yarn scrap is mixed with recycled office paper and used in packaging and hang tags.[7]

[7]For more information, please visit: https://tonle.com/pages/zero-waste.

- The service model stops offering a finished product, moving away from the regular purchase and sale practice, to offer a service. MUD Jeans offers customers the possibility of mending or recycling jeans as well as jeans lease service for a USD 7.5 monthly fee. After a year, customers can return the product or switch to a new pair. Then, the brand decides whether to send it as a vintage pair of jeans or to recycle the raw material, mixed with other textiles. They use organic cotton grown on wetlands, and they also customise buttons so that users can remove them for other purposes after throwing the jeans away. Unlike most denim brands, MUD does not use leather labels.[8]
- A longer lifespan means to design for durable products, while educating consumers to make more durable purchases. An example of this is Filippa K, a brand that looks for extending product life, offering the possibility of mending and remanufacturing garments. It encourages consumers to extend product life while making some money using the second-hand store, thus favouring the circular economy. Clothes are owned by the original buyers and remain in their hands until a new buyer or user appears. If the garment is sold, the original owner gets a profit. After a while, unsold items are donated [14].
- Residual value recovery refers to making the most of resources, recovering, and reprocessing raw materials to put them back on track. For example, company OrangeFiber creates textiles from orange waste, envisioning new uses for an organic material, which would, otherwise, be disposed of.[9]

In the textile industry, interface is one of the businesses that have implemented the four strategies. After reading Paul Hawken's book titled "The Ecology of Commerce," brand founder Ray Anderson realised that the operation of his company was not really environmentally friendly and that this type of operation would destroy the possibilities for future generations. Hence, in 1996, Anderson set the 2020 goals, radically redesigning the company's production system. Now, his goal is to be Net Zero, prioritising environmental protection.

The challenge is to migrate from a resource-mining and polluting model to an Earth life protective model. Anderson—a truly innovative and disruptive person, as well as a pioneer in the circular economy—succeeded in continuing with his company business: office carpet-tiles. But, now he offers a service. He does not sell but leases products. The company is engaged not only in carpet production, but also in carpet installation and subsequent collection. Therefore, materials re-enter the production process, using always renewable energies for processing.

Interface made this possible because its founder and CEO back then believed that sustainability was the only choice for development and growth. Ray Anderson's deep values enabled this shift required by the company, helping employees, shareholders, and users trust and embrace these principles.

---

[8]For more information, please visit: https://mudjeans.eu/.

[9]Source: Website: http://www.orangefiber.it/home/ Accessed: 15 December 2018.

4.1 Some Examples

As mentioned above, the textile industry is featured by long production chains. The following analysis describes different brands that are part of the circular economy and disruptive business models in the chain stages they operate. The circular economy is not only regenerative, but also collaborative and systemic. To ensure a fully efficient strategy, circular practices are needed in all the stages of the chain.

4.1.1 Excess Materials Exchange[10]

Relocating one business scrap to become another business input is, no doubt, to put in place a circular economy system. This company uses a digital platform to collect and allow for excess materials exchange for any other material or product needed.

Businesses can share and exchange all kinds of excess materials—from raw materials to finished products. This platform is a safe channel to trade these second-hand materials. As businesses usually pay for waste disposal, this practice offers a revenue, while maximising the value of resources and raw materials. Likewise, EME calculates the environmental and social impact of transactions and provides supporting reports. It provides a tool for attaining business sustainability goals.

Figure 7 illustrates the circular economy model suggested by this platform. It shows the relation and collaboration between businesses. In other words, it depicts how one business' waste becomes another business' raw material (Fig. 7).

On the basis of the above Tonelli and Cristoni's [27] principles, it could be said that Excess Materials Exchange offers the service of matching sellers to buyers. Moreover, it carries no product stock, as it is based on supply and demand, while reducing business environmental damage. This is the reason why it embraces Net Zero principle: to extend materials life by repurposing them as resources for another business, thus recovering product residual value.

4.1.2 Tejidos Royo[11]

Further down the production chain, some companies spin, weave, and manufacture textiles. Tejidos ROYO (TR) is a company engaged in the manufacture of high-quality denim. Founded in Spain in 1903, it exports fabrics to over 30 countries. It offers spinning, indigo dyeing, weaving, piece dyeing, and special finishing services. Sustainability and the circular model are part of this business DNA, which covers all three levels: social, economic, and environmental. Therefore, this should be present throughout the stages of the production chain.

---

[10]Source: Website: http://excessmaterialsexchange.com Accessed: 12 December 2018 and corporate documents.

[11]Source: website: http://www.tejidosroyo.com Accessed: 9 December 2018 and corporate documents.

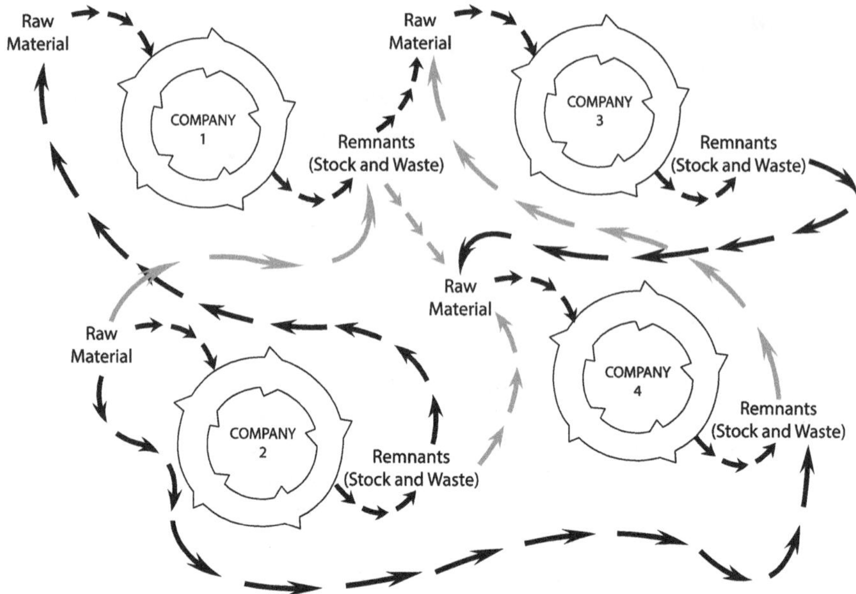

Fig. 7 Excess materials exchange circular economy model. *Source* Prepared by the authors

Denim is regarded as a highly polluting product in the textile industry. On the one hand, the raw material generally used is cotton—referred to as "the thirsty crop" for the amount of water needed to grow; and, on the other, indigo dyeing and finishing processes use highly polluting chemicals that are harmful to both the health and the environment—not to mention that they are also water-intensive processes.

TR develops quality products that ensure differentiation and exclusivity. They produce classic denim, and other highly sophisticated products and finishings. They have different certifications and endorsements to their quality and environmental and social concerns. These include, among others, the Oeko-Tex Standard 100 and STeP Certificate[12]; GRS[13] and OCS[14] Certificate by ICEA[15]; BCI[16] and Textile Exchange Member; Signatory of the UN Global Compact; Member of the AMFORI Audit Social Programme (BSCI[17]).

[12]Oeko-Tex is the most widely used environmental standard on a global basis. It belongs to the International Association for Research and Testing in the Field of Textile Ecology, based in Switzerland. Mowbray and Davids [21].

[13]Global Recycling Standard.

[14]Organic Cotton Standard.

[15]Institute for Ethic and Environmental Certification.

[16]Better Cotton Initiative promotes a fairer industry by reducing the environmental and social impact caused by cotton growing. Source: Mowbray and Davids [21].

[17]Business Social Compliance Initiative is a non-profit organisation based in Brussels that seeks to achieve and promote corporate social responsibility. Source: Mowbray and Davids [21].

They listen to their customers and create alliances, mostly basing their work on the design and development of strategies and products to meet their customers' needs, while respecting the environment and society. To ensure all this, they control product development in every step of the production chain, manufacturing their fabrics at their two plants located in Valencia. With emphasis on the importance of transparency, they offer product and material traceability throughout the entire production chain.

Some of the regenerative and collaborative projects that they promoted are:

- Refibra™ is a joint project by TR and Lenzing, a denim capsule collection designed by Adriano Goldschmed. It is the first cellulose fibre derived from post-industrial waste. It reduces the need to extract virgin raw materials, as it uses fabric remnants from eight global companies, which are processed by Tejidos ROYO. This is a new alternative to the well-known TENCEL® fibre.[18]
- Well in line with the above project, in 2014, TEJIDOS ROYO and MUD Jeans created a fabric made out of recycled denim fibres. They used both post-consumer and pre-consumer denim, considering the social, environmental, and economic aspects during product development.
- Back in 2009, the company created a new imitation leather finish for denim, offering a "different," high-quality product, which is, above all, an alternative to animal slaughter.

This case also represents the Net Zero principle, in an attempt to reduce textile manufacture environmental impact. In turn, it recovers the residual value of fibres and extends their life.

4.1.3 Rapanui[19]

Once the textile (fabric) is produced, it is sold and, eventually, garments are manufactured. An example of this is the British brand Rapanui, which tries to break the rules by implementing a more sustainable way of production and consumption.

In this regard, Rapanui seeks to protect both the environment and society from agriculture, using GOTS certified organic cotton.[20] This cotton is grown free of fertilisers or pesticides, using manure to keep the soil fertile. This favours biodiversity, generating softer and environmentally friendly materials, while taking care of the surrounding communities by reducing crop toxicity.[21] This type of cotton is grown in the wetlands of northern India, where monsoons provide the water required by the crop.

[18] Company MUD is working with this fabric, while delivering used garments and textile remnants to generate this fibre.

[19] Source: Website: https://rapanuiclothing.com. Accessed: 8 December 2018.

[20] Cotton seed remnants are pressed with other ingredients and used for animal feed, thus disposing of cotton processing waste and residues.

[21] Both to ensure transparency and to bring producers and employees closer to users, Rapanui features every person responsible for the different cotton processing and textile creation stages on its website.

The company pays fair, government-endorsed salaries to every employee, and those who deliver the best bundles receive an economic reward.

Products are manufactured in a factory[22] where the spinning, dyeing, weaving, and sewing operations take place. It is located in northern India, which is also the source of raw materials. Therefore, they can control the entire fabric and garment production process. This ensures improved quality control, while reducing transport costs and allocating those funds to employees' salaries and working conditions.

The factory is audited and controlled as per sustainability standards and presents itself as a socially responsible, SA8000 Certified (Social Accountability audit) and GOTS Certified company.

Given the water problem, and since the dyeing process is one of the biggest pollutants of this industry, the company dyes fabric using a recirculation process in which water is filtered, cleaned and reused in a closed loop.

As to shipping, all the products are carried by water. Products manufactured in India are plain, print-free textiles, and the printing process takes place in the UK. This is how they control stock, as it takes almost 10 weeks for products to get from India to England, and the company produces on demand. This way, they seek to reduce potential emissions and pollution caused by air transport.

As mentioned above, the company uses renewable energies. Moreover, and to reduce chemical negative impact, they developed special printing technologies. In turn, consistent with the circular model and open-source principles, the company shares the development of these technologies.[23]

Therefore, as production is on demand, they reduce stock and product costs—as unsold products would become surplus stock. These resources can be invested in education, e.g., delivering workshops and seminars to train and give opportunities to people too.

They also encourage users to reconsider how they wash clothes for improved durability and reduced footprint. Seventy per cent of carbon emissions occur in the use stage, during washing and drying. This is why Rapanui asks customers to wash clothes with cool water and hang them out to dry. This will extend clothes life and reduce pollution.

In turn, they aim at remanufacture based on the philosophy that products should not end up in a landfill or as waste at the end of their life cycle. This company is developing alternative fibres based on their own old T-shirts. Consumers can return their old T-shirts to backtorapanui.com in exchange for store credits.

The analysis of this case shows how the company can adjust its production model and, above all, redirect its business towards a circular economy system. Figure 8 illustrates the circularity of Rapanui's production chain.

In line with Tonelli and Cristoni's [27] circular economy principles, it could be asserted that Rapanui pursues the Net Zero principle, trying to reduce its environmental impact while extending product life and recovering residual value.

---

[22]They use renewable energy. In the UK, they have a solar energy plant. In India, the factory is fitted with two wind turbines.

[23]Information posted on: Teemill.com.

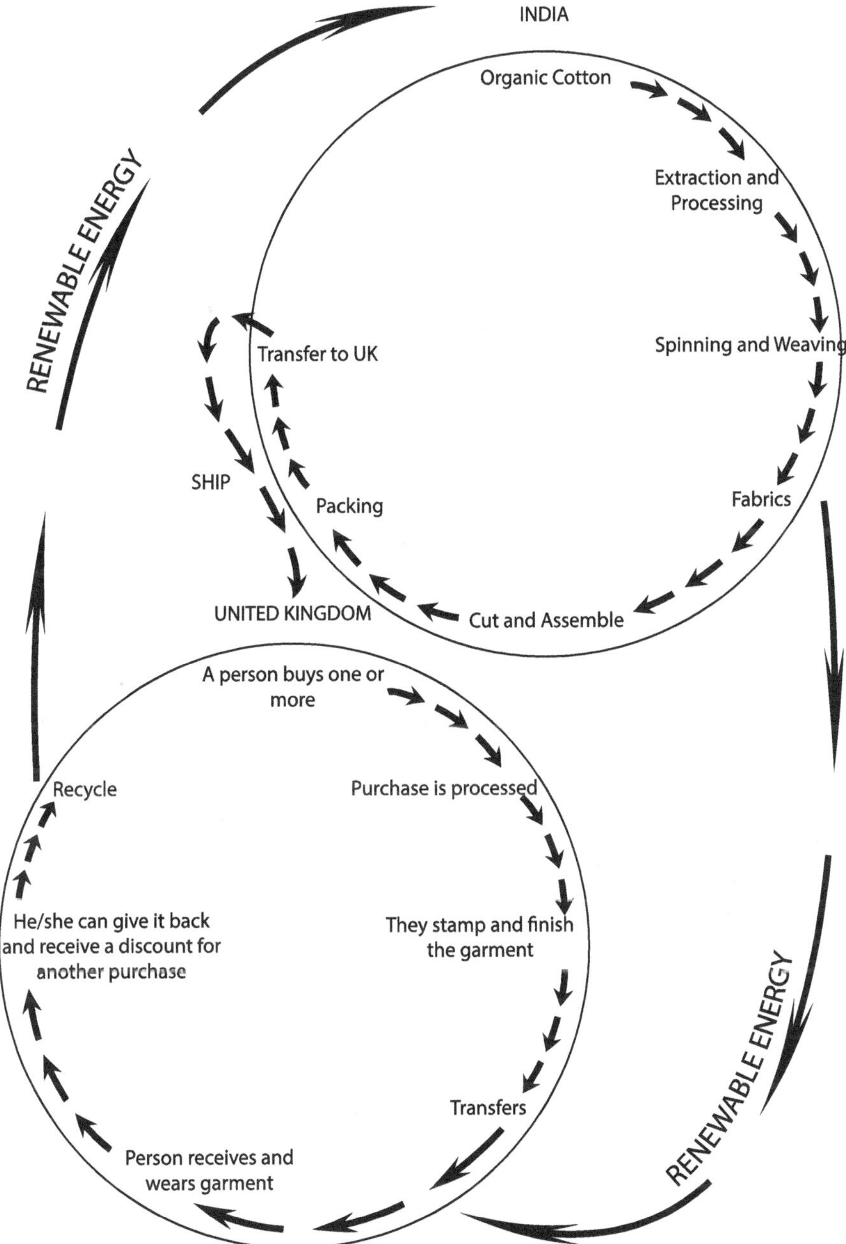

Fig. 8 Rapanui circular economy model. *Source* Prepared by the authors

Moreover, the company emphasises that education is vital to successfully embrace the circular economy.

4.1.4 Dutchawearness[24]

Dutchawearness is a Dutch company featured by textile industry innovation that pursues fully circular processes. It presents the circular economy as an alternative to sustainable development, turning waste into resources.

It poses transparency and co-creation as the key, since collaborating companies, designers, retailers, and users should lead change.

This company manufactures workwear created and designed to be reused, offering even circular alternatives to used garments that cannot be fully recycled. For a fixed monthly fee, companies can hire the lease service and, at the end of uniform life, if they are worn out or no longer needed, they can be returned. Then, materials re-enter the production cycle and the company refunds the money.

Moreover, they introduced Infinity DA Inside, a fully recyclable fabric. This fabric saves 95% of water and 63% of CO_2, while eliminating waste. Infinity textiles are available for a wide range of products, different types of workwear and different sectors. It is made of 100% high-quality polyester, with sufficient viscosity for recycling. It can re-enter the production cycle up to eight times, offering excellent recycling results.

Figure 9 shows the business production and recycling method. Note that, in every stage of this process, products can be traced using a code to check production chain status in real time.

First, polyester is shredded [7] (see Fig. 9) and melted [8]. Then, follows the spinning process (1), continuing with the weaving process, in which the fabric is manufactured (2). After customer interviews to find out their needs, the project is analysed (3). The next step is to review the information and prepare the order (4), which will be sent to production (5). Then, the product is customised (6) based on the customer's needs and, finally, it is delivered to the customer (7) who wears the product for an indefinite time and returns it. In this case, the company refunds the money as if the customer has paid a lease for garment use. Products are then collected (8) and re-enter to the production chain.

Likewise, the company has the so-called Cliff products, that is, another type of products other than Infinity fabric, which also re-enter the production process post-consumer use. These products are turned into recyclable materials and composites.[25]

In order to monitor materials and avoid delays in the production and supply chain, the company keeps a close track of product location in real time. To close the chains, they use a tool called CCMS—Circular Content Management System—a tracking

[24]Source: Website: http://dutchawearness.com. Access: 10 December 2018 and corporate documents.

[25]The firm has a separate website for these products: www.cliffstore.nl.

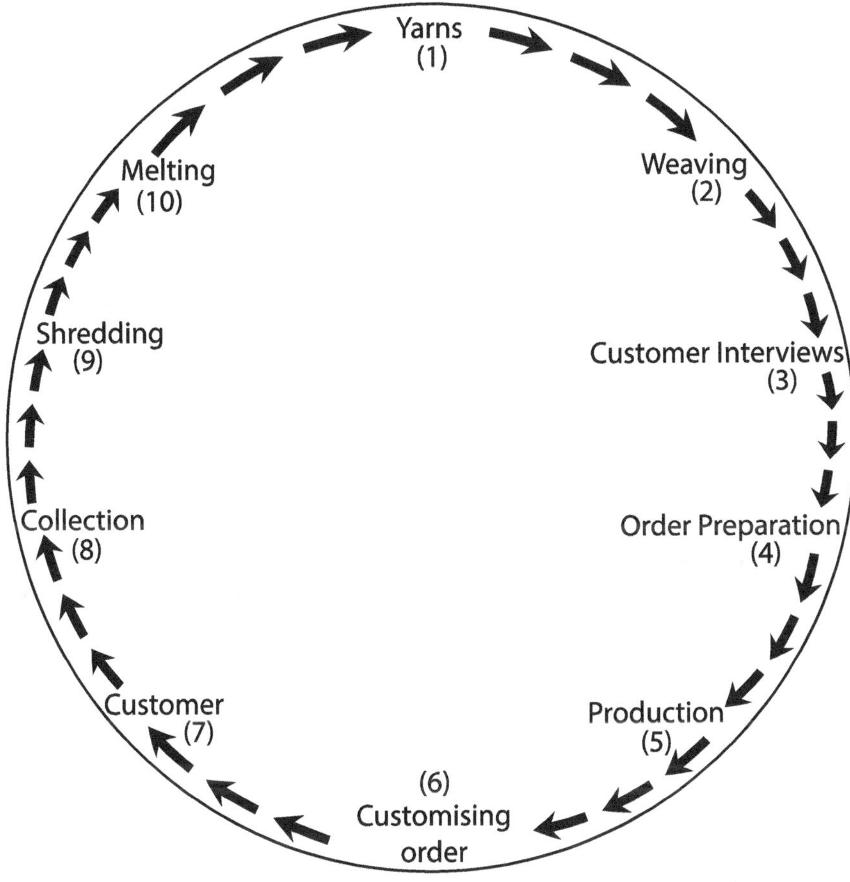

Fig. 9 Dutchawearness circular economy model. *Source* Prepared by the authors

and tracing system shared by all the supply chain partners. For tracking and reuse availability purposes, both raw materials and products are given unique bar codes.

With innovation and reverse cycle design, they work in line with the zero environmental impact principle, make a commitment, and create long-lasting customer relationships, while adjusting and developing new technologies and materials.

4.1.5 Stylelend[26]

Clothes servitisation gets customers involved in the circular economy, resulting in reduced production and resource use. Stylelend is a platform that seeks to match customers' needs to other people's potential.

In this case, you list certain garments and accessories from your wardrobe and another person rents them. Eighty per cent of the income goes to the person who lists the garments, and the platform keeps the remaining 20%.

In view of the current problems faced by the textile and fashion industry, there is no stock, unnecessary transport, or storage, and garments are put back on track. The platform offers a safe way to lend and rent garments: if the garment gets lost or spoiled, the platform replaces it or compensates the lender for its market value. Suggested prices are 5–10% of retail price. Garments are shipped by private courier and, when the item is back, it should be washed to keep it ready for the next order.

The company requires garments to be in very good conditions and manufactured as per ethical standards. For instance, it specifies the brands that could be listed, and those which could not be listed or even asked about.

Therefore, it promotes responsible consumption and, above all, it helps match a garment kept in your closet to somebody who needs it, making it flow and mobilising energy.

5 Analysis, Conclusions, Towards 2030…

After delving into the sustainable development objectives, circularity is evidently a clear example of SDG 12.

The figure below shows the main aspects of Sustainable Development Goal 12 based on the five businesses analysed above. The basic cores of this objective are listed in the centre with a number, and companies are listed in circles on both sides, with numbers inside the circles to show their relation to SDG 12 (Fig. 10).

In all the cases, two or more of the key aspects of these circular economy models are already implemented. Combinations are endless and, as said before, this is about thinking out of the box and figuring out new horizons.

Based on the above key aspects and problems of the current industry, it is imperative to find new ways of production and consumption. Along this line, the circular economy provides for a path of personal and industrial rediscovery, based on the continuous flow of objects while keeping their value. It is a question of organisation and thinking differently.

The textile industry could be the pioneer in environmental protection, not only because of its scope and impact on development, but also because clothes are worn by people. People wear clothes from different brands and may endorse their causes,

---

[26] Source: Website: https://stylelend.com/how-it-works. Accessed: 11 December 2018 and corporate documents.

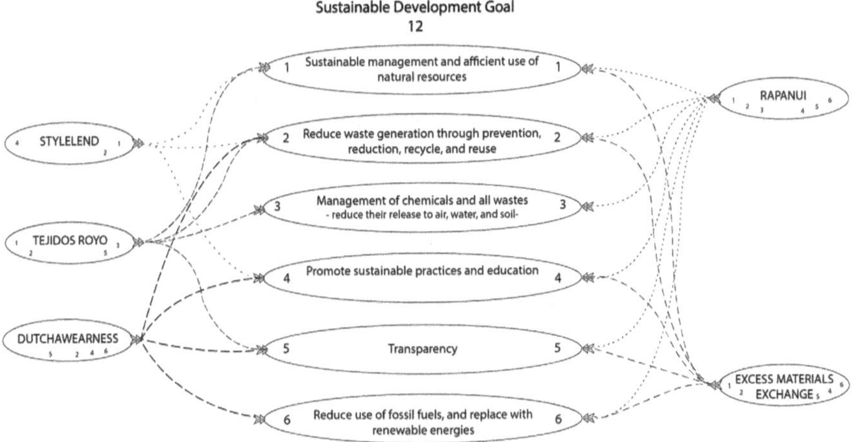

Fig. 10 SDG 12: key aspects and their relation to brands. *Source* Prepared by the authors

wearing an ideal, a value, a way of being, and doing in the world. Today, it is vital to reorganise resources and to educate, both inside and outside the company.

Sustainable Development Objectives set the course: the 2030 Agenda. The only way to achieve them is by means of collaborative, regenerative projects that, above all, respect human beings and the planet. If we intend to continue living on this planet for many generations to come, companies need to change their priorities and focus. The current business models are usually focused on economic revenue and profit maximisation. Now, it is time to prioritise environmental protection and social capital. While they do not exclude each other, development can only be achieved through environmental and social protection because, otherwise, as mentioned above, we would destroy the future of both the industry and life on Earth.

In line with this, businesses should reorganise their resources, assets, capital, roles and potentials, seeking nothing but to implement the circular economy principles at the core of the firm. This will ensure not only the future of the company and the achievement of the Sustainable Development Objectives, but also the future of life on Earth.

References

1. Anderson R (1998) Mid-course correction, toward a sustainable enterprise: the interface model. Green Publishing Company, USA
2. Benyns JM (1997) Biomimicry: innovation inspired by nature. Harper Perennia P, New York
3. Charter M (2019) Designing for the circular economy. Routledge, London and New York
4. Fletcher K (2012) Durability, fashion, sustainability: the processes and practices of use. Fashion Pract 4(2):221–238

5. Fuertes F, Goyburu ML (2004) El Perfil sobre las Comunicaciones del Progreso en Argentina—Qué Comunican las Empresas del Pacto Global?. Global Compact Office in Argentina, Buenos Aires
6. Fuertes F (2005) El Rol de las Empresas en el Desarrollo. Paper presented during the first activity of the Learning Lab at the Base of the Pyramid, Buenos Aires, Sept 2005
7. Galí JM (2013) Consumicidio, Ensayo sobre el consumo (in) sostenible. Omnia Books, Columbia
8. Gardetti MA (2017) Textiles y moda ¿Qué es ser sustentable?. LID, Buenos Aires
9. Gardetti MA, Delgado ML (2018) Vestir un mundo sostenible: la moda de ser humanos en una industria polémica. LID, Buenos Aires
10. Gardetti MA (2019) Introduction and the concept of circular economy. In: Muthu SS (ed) Circular economy in textiles and apparel. Woodhead publishing, United Kingdom, pp 1–12
11. Gardetti MA (2005) Objetivos del Milenio y Pacto Global. Lecture delivered within the framework of the First International Seminar on Corporate Social Responsibility F'Responsabilidad Social Empresaria- Desafíos y Oportunidades: Hacia un Pacto Global en el Agro'. Facultad de Agronomía, UBA and Corporación del Mercado Central de Buenos Aires, Buenos Aires, 2005
12. Graedel TE, Allenby BR (1995) Industrial ecology. Prentice-Hall, Upper Saddle River
13. Hethorn J, Ulasewicz C (2015) Production and economic processes in the global economy—introduction to section II. In: Hethorn J, Ulasewicz C (eds) Sustainable fashion, what's next?. Bloomsbury, New York, pp 129–131
14. Kant Hvass K (2015) The Filippa K Story. In: Hethorn J, Ulasewicz C (eds) Sustainable fashion, what's next?. Bloomsbury, New York, pp 124–128
15. Lacy P, Rutqvist J (2015) Waste to health: the circular economy advantage. Palgrave Macmillan, London
16. Lehmann M et al (2017) Pulse of the fashion industry 2017. Global Fashion Agenda and the Boston Consulting Group
17. Mao J et al (2016) Circular economy and sustainable development enterprises. Springer Nature, Singapore
18. McDonough W, Braungart M (2002) Cradle to cradle, remaking the way we make things. North Point Press, New York
19. Mollison B (1988) Permaculture: a designer's manual. Tagari Publications, Tyalzum
20. Morlet A et al (2017) A new textiles economy: redesigning fashion's future. Ellen MacArthur Foundation
21. Mowbray J, Davids H (2010) Eco textile labelling guide. MCL GLOBAL, West Yorkshire
22. Nelson J, Prescott D (2003) Business and the millennium development goals—a framework for action. United Nations Development Programme and International Business Leaders Forum, New York
23. Niinimaki K (2017) Fashion in a circular economy. In: Henninger CE et al (eds) Sustainability in fashion. Palgrave Macmillan, USA, pp 151–171
24. Papanek V (1971) Design for the real world: human ecology and social change. Phanteon Books, New York
25. Radhakrishnan S (2017) Denim recycling. In: Muthu SS (ed) Textiles and clothing sustainability. Springer, Singapore, pp 79–124
26. Rutter C, Armstrong K, Blazquez Cano M (2017) The epiphanic sustainable fast fashion epoch. In: Henninger CE et al (eds) Sustainability in fashion. Palgrave Macmillan, USA, pp 11–31
27. Tonelli M, Cristoni N (2019) Strategic management and the circular economy. Routledge, New York
28. Vadicherla T et al (2017) Fashion renovation via upcycling. In: Muthu SS (ed) Textiles and clothing sustainability. Springer, Singapore, pp 1–54
29. Webster K (2017) The circular economy a wealth of flows. Ellen McArthur Foundation Publishing, London
30. Weetman C (2017) A circular economy handbook for business and supply chains, Kogan Page
31. Zátonyi M (1990) Una Estética: del arte y el diseño de imagen y sonido. Nobuko, Buenos Aires

Flax Fibre Extraction to Fashion Products Leading Towards Sustainable Goals

Sanjoy Debnath

Abstract Among the different natural fibres, flax is the oldest fibre from plant source used during the early human civilization. The use of the fibre is documented since ancient Egyptian era. This fibre is normally grown in different parts of Europe, America and Asian countries. Water is one of the important resources used in every process starting from cultivation to processing. Recycling/reuse/minimizing the use of water may be one approach towards sustainable development. On the other hand, reducing the use of chemicals/ natural resources reduces environmental load. This chapter covers to some extent about cultivation and extraction of the fibre and its further processing into yarn and fabric up to fashion garments. It will also touch on the aspects like reuse and bio-disposal.

Keywords Flax fibre · Processing of flax · Flax textiles · Reuse and sustainability

1 Introduction—*History of Flax Fibre, Pertaining to Today's Scenario*

The advancement of human civilization improves life protection and comfort. Initially flax fibre was used as clothing material but later on it was used for application in other diversified areas and technical textiles. A very less information is available when exactly human started covering tree barks or animal skin to protect body from nature and gradually shifted to textile fibrous material as clothing in the prehistoric times. With progress in civilization and rising necessity men started using natural fibre, as hand made felt like material which were suitable in wrapping to protect and cover their body. Further, with time, natural fibres from plant and animal sources were used to prepare textile materials in the form of yarns and fabrics used as protective fabric. Based on the records, plant fibres like flax, cotton, wool and nettle are

S. Debnath (✉)
Mechanical Processing Division, ICAR–National Institute of Natural Fibre Engineering and Technology, 12, Regent Park, Kolkata 700040, India
e-mail: sanjoydebnath@yahoo.com; sanjoydebnat@hotmail.com

M. A. Gardetti and S. S. Muthu (eds.), *The UN Sustainable Development Goals for the Textile and Fashion Industry*, Textile Science and Clothing Technology,
https://doi.org/10.1007/978-981-13-8787-6_3

some of the ancient fibres have used during those days. It is found that even during 2325 BC, the Egyptian used flax fabric to wrap the mummy. Literatures revealed that in ancient days, various parts of plants (root, leaf, stem, flower, etc.) and minerals to dye/colour the textile materials. Slowly, with the passage of time, the look and appearance of the textile material got importance, and human started using extraction of natural dye from various parts of plants to colour and decorate the textiles. Hence, this implies that, during the Egyptian pyramid era, flax was cultivated for the extraction of fibre to prepare textiles.

The flax or linseed (*Linum usitatissimum*) plant belongs to genus *Linum* and *Linaceae* family of flax/linseed. The flax fibre is extracted from the bast of the flax/linseed stem of the plant. It is a dual-purpose crop and is annually renewable, popularly cultivated in different parts of the world as food and fibre crop. However, the fibre/textiles made from flax are known in the Western countries as *linen* [5]. This fibre is traditionally grown to be used for bedsheets, underclothes and table linen, etc. On the other side, the flax seed is popular to extract oil, popularly known as linseed oil used in the industry. This extracted flaxseed oil is used extensively in industrial applications, like wood preservation, concrete preparation, paint and varnish preparation. This chapter mostly restricts about different aspects of flax fibre. Basu and Datta [5], have elaborated an overall view and its potentiality of specifically Indian flax fibre. They covered various aspects of cultivation to the final product, different properties of flax fibre and compared its major tensile properties with various popularly known natural and synthetic fibres.

The development in science and gradual intervention of technologies and synthetic fibres from petroleum by-products are being developed in the early 1930s and are commercially popular after few decades. Gradually, different types of such synthetic fibres are synthesized, and a wide variety of fibres is popularly used today. Most of these fibres are not biodegradable during disposal, causing nuisance to environment, and burning causes black smoke and hazardous fumes, which again pollutes the air. Understanding these issues, since last decade, the natural fibres again gained popularity. In the plant fibre kingdom, flax plays an important role next to cotton as far as clothing textile is concerned. Many new textile industries are now diverting and increasing the production of linen materials. Linen is regaining its importance in the fashion textiles, starting from clothing to home décor from linen fibre, and waste/tow flax is being popularly used to the extent of industrial textiles. Even the world's biggest online wholesaler as well as retailer deals with a gamut of flax products. Today, though flax is being cultivated in various parts of the country, it is today considered as a fashion product.

Today, in many countries, this linen/flax is cultivated not only as cash crop but also generates good revenue for the country's growth. In other words, the country's economy depends on this cultivation of this crop to some extent. As far as the degradability is concerned, since it is annually renewable natural fibre from plant origin, majority of cellulose base, it can be decomposed easily without causing environmental hazards.

2 Cultivation, Extraction and Processing of Flax Fibre—*A Total Value Chain Approach and Sustainable Goals*

Among the different ligno-cellulosic fibres, cultivation of flax is most oldest plant fibre used in wrapping mummies in pyramids as found from history. Most of the soil types are suitable for cultivation of flax fibre. However, among the soil types, alluvial with deep loams having larger content of organic matters are the most suitable for this fibre crop cultivation. Previously, flax was mostly cultivated little above the waterline in the bogs, along with cranberry plants. However, hard clay or dry sandy soils are most unsuitable for this flax cultivation. During farming, it requires less amount of fertilizer as well as fewer amounts of pesticides. After sowing of flaxseeds, within eight weeks, the height of the plant becomes 10–15 cm, and it grows everyday at a very fast rate until it reaches to the optimum height, i.e. between 70 and 80 cm, within 50 days.

2.1 A Total Value Chain Approach of Flax Fibre

All over the world, flax is cultivated mainly for two purposes, flax oil (linseed oil) and flax fibre. Other than this, the broken sticks/woody stem parts have good potential for the preparation of particle board manufacturing. In general, for fibre purpose, flax plant is harvested around 100–130 days of plant age. Normally, by this time flowers bloom and capsule formation takes place. Under this condition, the plants are quite green in colour, and the seeds present in the plant are immature for germination. In order to produce viable seeds, for oilseed or seed for nest cultivation, plants are allowed to grow and mature over 150–160 days. However, the fibres present in the bark become unsuitable when the plant and capsules turn yellow in colour and start to split. On the contrary, these seeds will be suitable for seed production. In Europe, combined harvesters are used to cut the crop ends or the whole plant. Further, the plants are dried and the seeds are extracted. Many times, the farmers do not get appropriate price of fibre as it contains weed, which is present in the flax field and harvested along with flax plant using combined harvester. Due to some enforcing situation, if the plants are not harvested in time, the crops are burned in the field, rather allowed to decompose as the stocks are very hard and take longer time to decompose. Instead of burning, these flax straw/stocks are collected from farm and sold for usage in shed preparation for farm animals, bio-fuel, etc. [1].

In developed countries, a specialized flax harvester usually harvests flax for fibre production. Usually, the combined harvester machine has a facility, wherein the flax plants are gripped using rubber belts about 20–25 cm above the ground, minimizes the collection/harvesting the weeds/grass along with the flax plants. As a result, the whole flax plant, only gripped by the rubber belt are uprooted from the soil by the combined harvester. This process eliminates weed/grass contamination in the fibre. Instead of cutting the plant, uprooting the flax plant gives longer fibre length. Later, the plants are allowed to dry, and then the flax seeds are collected prior to retting. Based on the climatic situation, the harvested and flax plants after removing the seeds, plants are allowed to remain on the ground for a period between two weeks and two months for proper retting. During this retting period, farmers usually turn upside down the plants for uniform retting of flax fibre. Under these circumstances, the pectin or the gummy matter present in the fibre, adhering the fibres to bind with the stick/woody part of the plant stock gets separated due to occurrence enzymatic/microbial action. Time-to-time examination of the fibres is made to find the optimum retting; once retting is over, the plants are dried and rolled over and formed in bundle, which is further stored by the farmers for fibre extraction from flax straw. Normally, dew as moistening/watery medium is used to grow microbes for retting of flax, and many of the times, this process of retting flax fibre is popularly known as dew retting of flax. The dew-retted fibre normally has more strength, darker brown/bluish in colour; on the contrary, the water-retted fibres have comparatively less strength, little coarser in nature and greenish-brown in colour. However, the water-retted fibre caused water pollution, and the time of retting is normally 72 h. In the case of water retting, after retting in water for 3 days, fibres are taken out from the water and allowed them to dry in hot sun on a concrete floor. During this time, the moisture present in the fibre will be dried up to a larger extent, and then the flax stocks are fed to flax decorticator machine. Normally, fibres stocks are allowed to pass through pairs of nylon serrated rollers, wherein the sticks got broken/fragmented (which are adhering to the fibre). The delivered material is further shaken in air to separate the broken sticks from the fibres. Further, a flax scutcher is used to remove all the woody stick/foreign particles. Fibre is collected in bundle of around 500 g, and bale press machine is used to form a bale of 150–290 kg depending upon the consumer requirement/company/country standards.

During cultivation of flax, different common disease (Anthracnose, Basal stem blight, Brown stem blight, Browning and stem break, Damping-off, root rot, seedling blight, Dieback, Pasmo, Rust, Stem mould and rot, Wilt) occurred by microbes for which many chemical pesticides used. In order to minimize the application of chemical pesticide, biological means/bio-pesticide may be consider as green technology to reduce environmental pollution.

In flax textile industry, the sorting and mixing of flax is the first step followed by processing through hackling, carding, series of drawing passages (six), roving and finally wet spinning [3, 7]. Sometimes, based on the requirement the flax roving is scoured and bleached using H_2O_2 (hydrogen peroxide) followed by wet spinning to obtain bleached flax yarn [6, 17]. In this process environmental friendly chemicals are used and the chemically treated products are safe to use. However, it should be ensured that, after the treatment, the chemicals/treated liquor must the treated prior to discharge to environment. Again some manufacturers use mineral/synthetic oil emulsion during processing prior to processing through hackling machine. Basu et al. [4] established bio-friendly conditioning agent on jute fibre spinning, which can be used for flax processing as far as the green processing is concerned. It was established that alkali-treated flax fibre beyond 30% in the blend of jute–viscose deteriorates

Fig. 1 Manual flax scutching of Indian flax fibre

the open-end spinning performance significantly [17]. The alkali treatment though generates the bulk and softness in the yarn structure but with the sacrifice of the fall in tensile strength. This blended yarn has wise possibilities to develop products from green fashion since the raw material is natural origin and biodegradable. There was a study on effect of tension in the dye bath on the dye uptake of flax fibre and found that there is significant reduction in dye uptake of flax fibre while dyeing under tension [6]. In the case of Indian flax fibre [10, 11, 12, 13, 14], an extensive trial has been made to explore the possibilities of preparation of fine as well as industrial flax yarn of higher strength. Indian flax fibre contains huge amount of sticks; hence, a suitable manual flax scutcher has been developed by them (Fig. 1). They used jute processing machinery for producing coarse and strong yarn in the tune of 13.9 cN/tex for 276 tex yarn. The line fibre has been spun in flax processing machinery followed by wet-spinning system. The yarn performance during spinning is satisfactory for 42 and 67 tex yarn which has good potential for use in the preparation of fashion textiles.

During manual scutching of flax fibres (Fig. 1), the broken stick particles entangled with the fibres and also some small amount of fibres adhering with stick particles are separated. This makes the flax fibre strand clean and free from foreign particles. The broken stick particles along with some entangled fibres, as waste generated during scutching can be used for particle board manufacturing. Bio-resin may be used to form bio-composite/bio-particle board. In most of the cases, urea formaldehyde-based resins are used for better bonding and cost-effectiveness. However, the synthetic chemical bonding is hindering the green sustainable processing [2].

2.2 Sustainable Goals of Flax Value Chain

To maintain the global standards of processing and environment aspects, there exist different sustainable goals. For maintaining a sustainable value chain of flax processing, there are as many as seventeen sustainable goals to address the global challenges faced by the industry, including corporate social responsibility (those related to poverty and inequality), climate, environmental degradation, prosperity, and finally peace and justice. All these goals are interconnected, and considering equal importance to all, it is very essential to achieve each goal and target by 2030. Out of which the following discussions are some of the important aspects of sustainable goals specifically for flax fibre value chain:

Clean water and sanitation: *Clean, accessible water for all is an essential part of the world we want to live in*

Unlike any textile processing, flax processing (scouring, bleaching, dyeing) involves clean water. To address this sustainable goal, industry must have water treatment plant and water recycling system. Another aspect of in this regard is reuse of dye liquor for the next dyeing to minimize the freshwater requirement. Further to reduce the water consumption, in wet spinning, recycling of water may lead to reduction in water consumption. Lastly, industry should have zero discharge wastewater treatment plant to avoid contamination of surrounding water bodies.

Affordable and clean energy: *Energy is central to nearly every major challenge and opportunity*

Use of unconventional energy is the only solution to address this issue. Industry should go for solar panel installation on the roof/shade of the plant, which can generate clean energy from sunlight. Also installation of windmill, if possible, may help in supplementing affordable clean energy generation issue in sustainable manner. Overall, these methods of adoption of energy generation are almost pollution free as well with low-carbon emission to environment.

Decent work and economic growth: *The sustainable economic growth, which require societies to create the conditions that allow people to have quality jobs*

This flax industry is one of the labour-intensive industries; hence, many people can be engaged in jobs. However, with the passage of time, people have started thinking of quality living. At the same time, the industry also has to adopt modern machine with lot of safety, comfort features along with ergonomic aspect of design. These in total may lead to clean, less noisy and good working atmosphere.

Responsible production and consumption: *Responsible Production and Consumption*

Industry has to cope up with latest machinery and systems to minimize the energy consumption; in other words, consumption of energy must be energy efficient. A matter is also concerned in this regard that the production-to-consumption ratio has to be low to address this sustainable goal issue. Installation of energy-efficient lighting system and energy-efficient motors minimizes the wastage of heat energy using proper insulator; and monitoring these on time may lead to increase in production of textiles with lesser consumption of energy.

Climate action: *Climate change is a global challenge that affects everyone, everywhere*

Climate change is one of the major issues in today's scenario. To address sustainable goals under this climate change, industry has to use more and more unconventional energy/green energy. This not only helps to reduce the climate change but also reduces the carbon footprint. Also, reduction of burning fuel in boiler by substituting with latest methods like solar panel heater, infrared heater and heat pumps with almost zero carbon emission and greenhouse gases helps in addressing the environmental issues. These are some of the suggestive methods that can be adopted by the industry to address this sustainable goal under the banner of climate change.

3 Fashion Textiles from Flax Fibre and Disposal

Normally, the fine flax fibre also known as line fibre is used to prepare summer cloths and fashion textiles [3, 16]. Apart from its aesthetic appeal, flax textiles have high as well as fast moisture-absorbing capacity in addition to dyeing capability in good shades. Based on the properties of the flax/linen fibre, fibre material gets swollen and gains its strength in wet condition [3]. Because of this special character of flax fibre, wet spinning rather than dry spinning method is preferred to spin the fine linen/flax yarn to achieve better spinning performance. In international market, a lot of pure linen/flax or blended materials are available. Some of the popular items like ramie-linen blended *Kurti* for women casual wear, viscose-cotton-linen blended trousers of different blends (50% viscose, 35% cotton, 15% linen; 65% viscose, 25% cotton, 10% linen) products marketed by Joanna Hope. Linen-cotton blended (55% linen, 45% cotton) trousers and short marketed by South Bay. Blazers made from single-breasted linen blend (55% linen, 45% cotton, lining material polyester) products marketed by Williams and Brown. Lightweight linen-cotton blended fabric (54% cotton, 46% linen) will keep the body cool and also stylish in the warm weather marketed by Black Level Jacamo. Linen-cotton mix (55% linen, 45% cotton) 3/4 pants marketed by South Bay Ladies fashion textiles from 100% linen material for bow decor straw braid for summer sun hat and summer clothing for women's wide brim for sun hat as well as wedding church made from 100% linen material, for using at sea beach marketed by Kentucky Derby. Ladies pleated crisscross fashion sexy wear for women's dress from 100% linen mateial and, linen-nylon-spandex fibre blended women garments like, slim blazer short jacket/linen blazer/ladies coat, ladies thong linen underwear/briefs and linen-nylon blended (85% linen, 15% nylon) women shocks established the fact that a good fashion market for flax-based textiles available internationally.

Flax is one of the oldest fibre plants and hence not much special attention required during the production of the yarn and fabric for flax fibre. Because of these reasons, many of these fibres like flax and cotton are being used since invention prior to much developments in science and technologies. Most of the plant fibres are extracted from different parts of the plant and throughout the process of preparation from fibre to fabric is sustainable and final products are biodegradable [9]. However, as the science and technology progresses, in the past 60 years, many synthetic/artificial fibres have been invented and are being popularized as the materials are mostly engineered as per the pre-requirement. After using these synthetic fibres for 40 years from its initial development, in last 15–20 years, special attention has been given to minimize the use of synthetic fibre, rather more attention has been given on the use of natural fibres at different possible applications areas and flax is one of the fibre components. Hence, flax fibre got special attention and the application of these fibres in apparel textile and technical textiles fields including fashion areas also increased gradually. To fulfil the industrial demand, more flax fibre are grown in limited area. This is due to reduction of agricultural land day-by-day because of expansion of urbanization. This demands more chemical fertilizers, hindering the future crop due to the loss of soil fertility in due course of time. However, with this improved productivity more importance has been given on the holistic approach on the sustainability of the production system.

Over a period of time, many cottage and small-scale industries are involved in manufacturing the flax and flax-based textile and fashion products using traditional/small-scale processing machinery. They use very minimal quantity of chemicals or sometimes use traditional plant extracts and natural materials to process and give value addition to the final product. Even some of the decentralized sectors are involving in producing fashion wear and flax/linen-based home textiles. With the increasing demand for flax textiles, last few decades, composite industries are coming forward to produce flax-based textile and many lifestyle products out of it. Worldwide, some of organized industries are concentrating on dealing on only natural fibre-based like flax for fashion and textile utility products starting from product development to promoting flax-based textile to the society. Even, the waste generated during flax spinning as flax tow fibre (flax hackling waste) is used for producing high strength technical textiles. Hence, considering the above future prospects of this fibre, it is the order of the day to concentrate more on the process and product out of flax and flax-based material towards their sustainable goals.

Unlike other natural fibres [8], flax can be renewable annually which may deplete natural resources and at one point of time natural imbalance will occur. Hence to address this holistically, one has to think and give importance (Fig. 2). It is found from Fig. 2 that except the availability of flax fibre, most other aspects like durability, colouration, disposability and cost end use opportunity parameters are in acceptable level for fibre. On these bases of important parameters, sustainable luxury textiles from flax/linen-based material can be considered. Graphical representation using polar diagram shown in Fig. 2 is conceptually proportionate to an approximate, as no actual industry standard/information is available in this regard. Apart from these

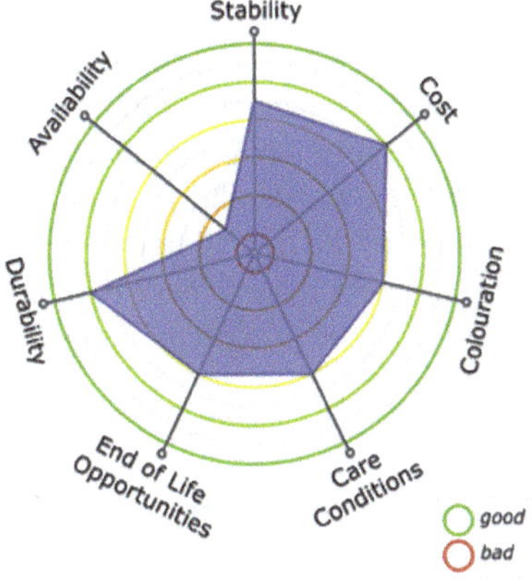

Fig. 2 Polar diagrammatic representation of comparative parameters of nettle fibre [8]

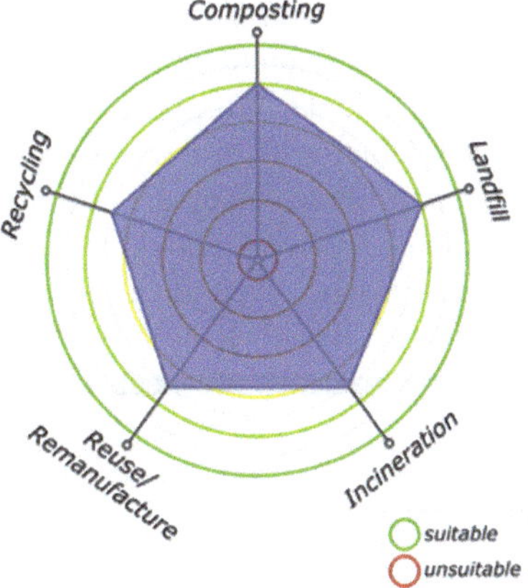

Fig. 3 Disposal of natural textiles [8]

factors, fabric construction, weight and blend composition with other fibres (for blended products) may influence the perceivable ranking to some extent.

Renewable fibre like flax after the successive usage, flax-based textiles, can be disposed off safely to nature (Fig. 2). Flax and flax natural fibre-based textiles may be considered in similar fashion like other natural/cellulosic fibre materials disposed after the end of usage. Some of the leading industries produce flax/flax cotton-based blended materials, which can be either recycled to form regenerated cellulosic fibres/products addressing the sustainability issues. Hence, there is a good prospect to recycle or re-manufacture the end used/disposed flax and flax-based textiles to generate cellulose feedstock as raw material for regenerated cellulosic products. Considering the biodegradable aspects, safe collection or disposal for composting after end usage can be a good solution for greener earth as depicted in Fig. 3 in the case of nettle fibre.

Water is one of the major sources in any agriculture/technological intervention. Water saving during agriculture can be attended using advance/controlled irrigation system and bio-mulching application. Although these interventions involve additional costs which is negligible if we consider the environmental and soil water conservation aspects. A significant amount of water is again required during chemical processing (scouring, bleaching, dyeing, etc.) of flax textiles. Optimum use of environmental safe chemicals, reuse of chemicals to a maximum extent and proper treatment of the treated liquor after chemical treatment prior to disposal may lead to save water resource. These may be a holistic approach, which leads to sustainable goals for future prospects in this flax industry.

4 Conclusions and Future Expectancy

Flax being an ancient natural fibre resource has promising potential for future fashion and handicraft industries [15]. The production of synthetic fibre with time will decline, and in future, time will come when there will be no such hydrocarbon material be available. With reference to these, there will be an end to synthetic fibres. Hence, judiciously, use of the water and energy will produce a significant amount of natural fibres, which are still not explored potentially. In the context of sustainable goals, the focus should be given on extensive application of flax-based textiles through product diversification emphasizing the eco-friendly aspects of processing technology. Based on the discussions made in this chapter, it can be summarized that there are immense prospects of exploring possibilities of flax-based textiles in the various textile applications (technical textiles). In order to popularizing the product made out of flax-based material, more popularizing through advertising platform (electronic as well as print media) is needed. Further, stress need to be given on using green processing chemicals or minimum use of chemicals with appropriate processing technologies to address the sustainable goals issue. Minimum use of hazardous chemicals and maximum use of natural resources during cultivation to product development will lead to green process and can meet sustainable goals. In addition, flax/linen fibre can be blended to a little extent with manmade/synthetic fibres to make the textile product sustainable. Overall, the people involved in producing these green fashion products from flax fibre will be remunerated, and a good amount of people will be benefited out of it apart from its environmental impacts. Finally, the industry must be equipped with the latest norms and systems to cope up with the sustainable goals.

References

1. Anonymous (2019) https://www.producer.com/2008/03/the-last-straw-nine-ways-to-handle-flax-straw/. Dated 13 Feb 2019
2. Banerejee PK (2001) Development of textile products for protection and enhancement of environment. Indian J Fibre Text Res 26(1 and 2):214–222

3. Basu G, Roy AN (2008) Blending of jute with different natural fibres. J Nat Fibers 4(4):13–29. https://doi.org/10.1080/15440470801893323
4. Basu G, De SS, Samanta AK (2009) Effect of bio-friendly conditioning agents on jute fibre spinning. Ind Crops Prod 29(1–2):281–288
5. Basu G, Datta M (2014) Potentiality of Indian flax. National Institute of Research on Jute and Allied Fibre Technology, Kolkata, India, pp 1–104
6. Chattopadhyay DP, Samanta AK, Nanda R, Thakur S (1999) Effect of caustic pretreatment at varying tension level on dyeing behavior of jute, flax and ramie. Indian J Fibre Text Res 24(1):74–77
7. Debnath S (2014) Machinery for fibre processing and latest developments in the area. In: Nag D, Ray DP (eds) Jute and allied fibres—processing and value addition. New Delhi Publishers, New Delhi, India, pp 111–120. ISBN: 978-93-81274-41-5 (Print)
8. Debnath S (2015) Chapter 3: Great potential of stinging nettle for sustainable textile and fashion. In: Gardetti MA, Muthu SS (eds) Handbook of sustainable luxury textiles and fashion, environmental footprints and eco-design of products and processes. Springer Science+Business Media, Singapore, pp 43–57. https://doi.org/10.1007/978-981-287-633-1_3
9. Debnath S (2016) Unexplored vegetable fibre in green fashion. In: Muthu SS, Gardetti MA (eds) Green fashion, environmental footprints and eco-design of products and processes. Springer Science+Business Media Singapore, pp 1–19. doi: 10.1007/978-981-10-0245-8_1.
10. Debnath S, Basu G (2017a) Extraction and processing of Indian flax fibre. In: Training manual of National Level Training Programme on Production and retting technology of Jute/Mesta/Ramie/Sunnhemp including other Related Aspects, Sponsored by National Food Security Mission (NFSM), Commercial Crops, Department of Agriculture & Co-operation, Ministry of Agriculture, Govt. of India from July 17–19, 2017, at ICAR-NIRJAFT, Kolkata, pp 75–79
11. Debnath S, Basu G (2017b) Processing of Indian flax fibre. In: Training manual of Exposure Visit-cum Training Programme on 'Innovative Agricultural Practices for Production & Processing of Jute & Allied Fibres', Sponsored by Block Farmers Advisory Committee (BFAC), Raghunathganj Block-I, Murshidabad, West Bengal from July 28–29, 2017, at ICAR-NIRJAFT, Kolkata, pp 44–49
12. Debnath S, Basu G (2017c) Prospects and processing of Indian flax fibre. In: Training manual of ICAR Sponsored Short Course on Recent Advancement in Processing Technologies for Value Addition of Jute and Allied Fibres, from December 11–20, 2017, at ICAR-NIRJAFT, Kolkata, pp 62–68
13. Debnath S, Basu G, Mishra L, Das R, Karmakar S (2018) Extraction and spinning of Indian flax fibre. In: Proceedings of National Seminar on Market Driven Innovation in Natural Fibres organized by The Indian Natural Fibre Society, February 22–23, 2018 at ICAR-National Institute of Research on Jute & Allied Fibre Technology, Kolkata, pp 33–39
14. Debnath S, Basu G (2018) Indian flax fibre—extraction and spinning. In: Training manual of National Level Training Programme on Production and retting technology of Jute/Mesta/Ramie/Sunnhemp including other Related Aspects, Sponsored by National Food Security Mission (NFSM), Commercial Crops, Department of Agriculture & Co-operation, Ministry of Agriculture, Govt. of India from July 23–25, 2018, at ICAR-NIRJAFT, Kolkata, pp 60–65
15. Dogan Y, Nedelcheva AM, Dragica OP, Padure IM (2008) Plants used in traditional handicrafts in several Balkan countries. Indian J Traditional Knowl 7(1):157–161
16. Sinclair R (2015) Textiles and fashion: materials, design and technology. Wood Head Publishing Limited, Cambridge, UK
17. Tyagi GK, Kaushik RCD, Dhamija S, Chattopadhyay DP (2000) Effect of alkali treatment on the mechanical properties of flax-viscose OE rotor spun yarns. Indian J Fibre Text Res 25(2):87–91

Sustainable Consumption and Production Patterns in Fashion

Shanthi Radhakrishnan

Abstract Fashion calls for constant change which urges consumers to indulge in purchasing garments to keep track with the latest trends. These practices result in large consumption patterns of clothing while the nonuse of fashion apparels is hidden truth. Clothes that are purchased with intense interest lie idle either in the wardrobe or in the landfill. The textile and apparel industry invest huge resources for the manufacture of garments to meet the demand of the consumers. Labels of top brands, retail houses, marketing divisions and advertising sectors add glamor to fashion to draw the attention of the consumer and manufacturers involved in fabric production and apparel development work towards high targets at the cost of environmental and social implications. The fashion industry is the backbone of textile and apparel production as all endeavors start with the design phase. Designers today have a different motto, and they work for long-lasting sustainable designs which in turn will promote sustainable consumption and production patterns among end users. This chapter will analyze the role of sustainable design development, the awareness of slow fashion and change in mind-set of the consumer to attain the 12th goal (Responsible Consumption and Production) of the UN Sustainable Development Goals.

1 Introduction

The world we see may look to have plenty of resources for every human being, but these resources are not endless for everyone to live how they feel and how they like. Mankind has used these resources for a long time, and there seems to be some constraints and limits to growth and development due to the large disparity in the distribution of resources. Serious injustice to natural ecosystems, water, and quality of air due to industrialization and exploitation has caused several impacts on the environment. Confusion may result due to theories like 'every action has an equal

S. Radhakrishnan (✉)
Department of Fashion Technology, Kumaraguru College of Technology, Coimbatore, India
e-mail: shanradkri@gmail.com

M. A. Gardetti and S. S. Muthu (eds.), *The UN Sustainable Development Goals for the Textile and Fashion Industry*, Textile Science and Clothing Technology,
https://doi.org/10.1007/978-981-13-8787-6_4

and opposite reaction'. Many people believe if one part of the globe is exploited, the other regions will become bountiful and generous in resources. However, 'endangered species' and 'vanishing indigenous cultures' are the signs of the depletion of resources due to the lack of concern for the environment.

Human beings have a great role to play with regard to Mother Earth, and there are many responsibilities that are existent in terms of our relation to every living phenomena and the environment around us. The root cause of the current environmental and social challenges can be rested on the dominant patterns of thought and emotional longings that move toward unsustainable contemporary lifestyles. A shift in mind and heart with care and respect for the living creatures can create a society which will flourish to become sustainable. Once this trend surges there will exist a consciousness that will engulf all to work toward a better future where people use natural resources minimally to save them for the future generations.

2 Predictions on Sustainability and Fashion for the Near Future

2.1 Transformation of Mass Market and Fast Fashion to Custom Made and Classics

The global apparel industry had been working tireless to change the fashion in retail stores in a fast manner (once in 15 days) leading to consumers making constant purchases to keep up with the latest trends. Wardrobes were overflowing and fresh new garments, not even worn once, may land up in waste bins as it was not in vogue. People have now become tired of mass made similar looking styles and are seeking new opportunities which have brought back the concept of tailor/custom made and bespoke clothing. Advertising and promotional activities have been spreading this message, e.g., Raymonds advertisement in Indian Television. The term 'Fashion on demand' [22] entails the customer to choose from an array of pre-designed options, styles, fabrics, colors, size and fit, etc., to cocreate the design using mix-and-match theory. The design output is also viewed by the customer and the choice is made. The order takes three weeks by which time the customer gets into an emotional bonding with the specially made outfit that makes him use it for a longer period when compared to the other outfits. This method removes the risk of overproduction, need for storage and warehousing, waste minimization, landfills and incineration, required use of virgin materials.

2.2 *Closing the Loop with Circular Fashion*

Circular fashion is popular among fashion conscious people. This movement is crossing all borders worldwide from America and Europe to Africa and Southeast Asia. A redesigning process is taking place in the overall business of retailers and brands and a change is creeping in from visions, strategies, procedures, and processes across the supply chain. Sustainable fibers and raw materials are being selected, testing is carried out to eliminate harmful substances, new collaborations are initiated, and special service packages are offered to the customers. Many programs have been initiated to spread the awareness of circular economy like Circular Textiles Program (2014) by Circular Economy, Fashion Positive Program [19] by Cradle to Cradle Product Innovation Institute, Textile Environment Design TED, a project at Chelsea College, University of Arts, London, 'Design for Redesign' by the Swedish School of Textiles, Sweden, doctoral dissertation by Dr. Kirsi Niinimaki on 'From Disposable to Sustainable', 'Close the Loop' a joint project of Plan C & Flanders Fashion Institute [10, 19, 50]. In 2017, Green Strategy has released the 'Circular Fashion Framework to discuss three basic principles—the meaning and definition of circular fashion, key principles to circular fashion, categories (8) of products and services of circular fashion.

2.3 *Compassionate Fashion—The Fashion Feel*

The most interesting concept emerging as a fashion trend is 'compassionate fashion'. The ethical and socially responsible aspect of the fashion movement is to respect, workers, animals and natural environment; the current change is the shift in attitude where the fashion industry is becoming compassionate about the impact of the industry across the supply chain. There seems to be a transition from consciousness to feeling real compassion for people, animals, and ecosystems. Slogans like 'Fur free Friday', 'I can't bear to see you in that fur', Cruelty of the Down Industry, Cool cruelty free over the knee Boots, Liberation from cruelty free culture, let them be free emphasize the change to compassion [45].

3 Fashion and Consumer Psychology

3.1 *Consumer Needs*

Design in the field of fashion is governed by attitudes, time and place considering customer needs and desires. Fashion is a code/ a tool/ a gizmo to read and understand society. The silhouette, form, and shape keep changing, but the function remains constant. The media help in showcasing and announcing the fashion statement. In the UK,

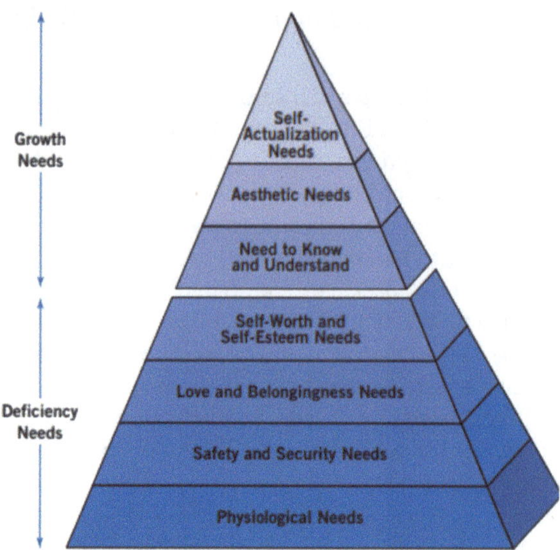

Fig. 1 Maslow's hierarchy of needs [36]

Vogue the fashion magazine sold 71,649 copies in the first half of the year 2018 while the Cosmopolitan sold 1,84,566 copies during the same time frame [49, 53]. Sociological and demographic factors like family, peers, neighbors, culture, social class, informal, and non-commercial sources influence consumers' choice of clothing.

Motivation encourages a consumer to choose and buy fashion products, and once the fulfillment is considerably high, the consecutive motivation will make them repeat the purchase [36]. According to Maslow, all needs are grouped as deficiency needs and growth needs as shown in Fig. 1. The deficiency needs can be easily fulfilled, but the growth needs are insatiable and they ask for more once one segment is satisfied. In the case of physiological and safety needs, function is the most important, e.g., purchase of a winter coat in the winter season; inflammability finish for children's nightwear [30, 36, 37]. After the physiological and safety needs come the love and belongingness needs where the consumer is impacted by family, peers, roles, and status, e.g., need of T-shirt by a teenager. From street fashion to haute couture, the whole fashion spectrum is affected by esteem needs and need for self-actualization. Esteem needs include the acceptance of the consumer by the others, maybe it can be coined as 'prestige' or 'social acceptance'. In the case of self-actualization [23, 54], the consumer moves to another stage and wants to showcase his creativity and individuality to stand apart from the others yet receive the awestruck acceptance of his social environment. Here the consumer maintains social uniqueness and image enhancement. These needs of the individual or consumer play an important role in the decision making and purchase pattern of the consumer.

3.2 Adoption of Fashion by Consumer

Fashion has become an integral part of one's life as it states his/her personality, culture confidence, and position in society. Conscious people cannot extricate themselves from fashion as clothes are communication devices. The adoption of fashion is based on frequency of purchase, magazines, and catalogues, readership, and online connectivity. The 'trickle across' theory is more apt for styles that gain adoption over time and across socioeconomic groups. Leaders and celebrities play an important role in the adoption of fashion. International stars like Madonna, Beyonce, and Sara Jessica Parker and influential role models for many customers and the styles offered and popularized in the market have a bearing on the promotional activities and the campaigns [24]. Brand awareness, fashion image, new trendsetting, and unique niche creativity are the functions of these shows and launches.

Peer pressure is an important stimulus to fashion purchase. According to Freud's theory of personality, outside awareness has a great influence on the purchasing behavior of the consumer. The influence on the choice of fashion is Personality which has been structured into three types. '**The id**' is unconscious impulsive nature which seeks fulfillment without any fashion consciousness. '**The superego**' is the individuals' context of the moral code of conduct. Here social consciousness is great and the need to satisfy a socially acceptable fashion is important. '**The ego**' controls the mind of the consumer in three levels—conscious, preconscious, and unconscious mind. According to this personality theory, consumers are driven by unconsciousness and are unaware of the reasons for buying fashion products. A special mention should be made on mass media for the contribution and control they have in shaping the minds of the consumers and for creating fashion awareness.

3.3 Changing Mind-Set of the Consumer

In the UK, between the years 1995 and 2005, the fashion consumption increased by one-third through the amount spent on clothing was 12% of the total household income when compared to 30% in the 1950s [3, 32, 40, 41]. This trend was noticed in all the Western countries and is attributed to cheap clothing, low prices, short life spans, and higher disposable income. Further, the study also highlighted the fact that almost half the garments shopped had not been used during the last one year, the estimate being 2.4 billion items [8] and these unused clothes are owned by young consumers (25–34 years). Another study in the Netherlands revealed that consumers kept their clothing for an average of 3 years 5 months of which the average times, it was worn is 44 days [20, 34]. Along with this trend rose another totally different tendency as seen in the denim jean which never faded from the minds of the consumers. People from all walks of life felt denim jeans were comfortable, durable, softening with age with the color acquiring a uniqueness that was different for different consumers. The indigo dye found on the top of the thread weared away as time passed causing the fabric to fade making it into a classic. Denim jeans also known as second skin, has a body affection, emotes romance, high spiritedness, dynamism, rebellion, character, carefree, hardworking, feeling of desperado, villainess.... the list goes on and in a nutshell; it reflects the lives of people [27, 46]. This illustrates that the same consumers can have different attitudes toward different apparels.

The present-day trend calls for sustainable consumption where durability and longer life of garment are the need made with artistically aging materials, exquisite design, and enduring style. Services for garments can include value additions like print over basic prints, redyeing, designing with dyes (batik, tie and dye) surface ornamentation (kantha work), repair, and upgrading that promote emotional bonding. In India, the traditional sarees are handed over to the next generation with great reverence and may be upcycled for the new user. Sustainable design should also include servicing, refurbishing, renovation, and transformation to increase the longevity and use of the product. Slowing down of consumption and introducing special unique features unlike mass production will add value to the goods fetching the price and profit. What the consumers need to do is purchase fewer, meaningful durable garments made from eco-friendly materials and processes bestowing longevity fueled by repair and restoration. Consumers experience a strong emotion when they are involved in consumption and purchasing situations; recreating different alternatives that can bring in such experiences and satisfaction in terms of sustainability rather than buying new fashion items.

Sustainability means different to different consumers. In a study conducted in Denmark about the attitudes of consumers on sustainable fashion, there seemed to be interesting results. About 30 participants were selected from online sources by theoretical sampling. The results gave insight into a lot of contradictions [9].

- The respondents felt that sustainable fashion cannot be coined together as they have associated sustainability with 'long term, slow, conscious, being stable about future' while fashion is all about 'speed, change, fast consumption'.
- These two have been used together for commercial interests and has a flavor of greenwashing.
- Secondly, sustainability was holistic and consistent and required attention on the entire life span of the garments. Environmental and social considerations are also to be taken while taking sustainable fashion but manufacturers and retailers talk about eco-friendly products without taking the entire sections, e.g., H & M have launched 'Conscious Collection', but they are facing serious litigations on the labor side. Some of the abuses include Abuse of workers in H & M and Gap factories, failure to ensure fair wages for factory workers, violation of labor rights in Uzbekistan and Bangladesh, female workers in Asian factories face sexual harassment, Myanmar factories employ workers under 14 years of age [5, 12, 18, 25, 26, 29].
- A lot of complex confusion and mistrust exist on these terms and consumers have become aware while shopping. This has led to different understanding of sustainable concept in fashion by different consumers.
- Respondents felt that sustainability is all about good intentions, but in the case of manufacturers, there was an overall sense of the concept without delving deep

into the same. They could not link profits and sales growth with sustainability as this would make all efforts futile.

- Consumption is the opposite of sustainability, as fashion is a fast model associated with cheap prices and overconsumption. Manufacturers project themselves to be sustainable, but they do not adopt it into their system. For example, H & M's Conscious Collection is almost 1% of their total turnover, and yet they call themselves a sustainable manufacturer/retailer. The current levels of purchasing are close to overconsumption. Most respondents felt that cheap prices and ever-changing assortments are the cause for thoughtless purchase which are stacked away or enter landfills as throwaways.
- Sustainable consumption brought about a feeling of satisfaction and well-being as they had contributed to the environment development by avoiding unnecessary buying. This helped them to accept their body image and enjoy a sense of freedom and relief from the thought of comparing to the idealized body image promoted by the fashion industry.

4 Sustainable Production Patterns

About 10% of the global CO_2 emissions, 20% of global wastewater, 24% insecticides and 11% pesticide use are caused by the production of fashion products. Efforts are being taken on a war footing to reinvent the industry, e.g., No Carbon dioxide initiative (No CO_2) for addressing the social and environmental footprint, eBay's Giving works, and Walmart Miracle Balloon Campaign. eBay's Giving works collects donations from buyers and sellers and extends its arms to new enrollment to broaden the support. Celebrities have been used in advertising, and this program has fetched around $521,000,000/—since 2003. Similarly, Walmart and Sams Club Miracle Balloon campaign has fetched donations of $380,000 [1, 16, 17, 51], indicating that people are willing to pay for a great cause. In the case of the No CO_2 program, diesel power used by factories will be replaced with clean energy, and the savings from this venture will be diverted to the factory workers. To create a circular economy reuse of fibers, living wages, organic cotton, and regenerative agriculture are essential, and people can be made to buy these products for longevity and sustainable consumption. So to attain sustainability in the fashion industries, support and cooperation from consumers will be extended and they will be willing to pay more for the same.

While choosing ethical clothing, the main criteria will be the minimal/ no use of toxic chemicals, lower use of land and water and reduction in GHGs. This cannot be achieved is the design of the product is not sustainable. Sustainable Production is based on sustainable design thinking and proto sampling should give priority to sustainable development in all phases of the life cycle of the product. In some cases, the industries have adopted measures to reduce and recycle materials by undertaking new technology and production processes. The raw material growth requires farming and cropping that reduces soil erosion by improved irrigation methods, reduced

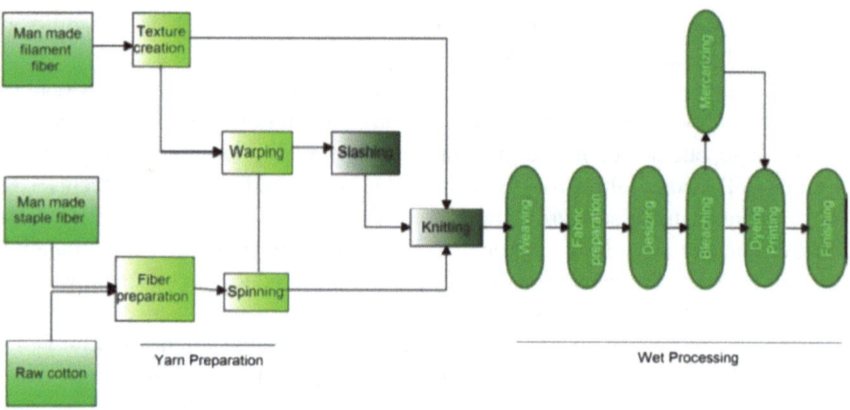

Fig. 2 Stages of Yarn preparation [4]

pesticides, and toxic chemicals, e.g., GMO cotton. It has been reported that the farming of cotton for one T-shirt requires (151.33 gms) one-third of a pound of toxic chemicals which includes pesticides, fertilizer, and defoliation chemicals [33].

Cotton is considered to be a natural fiber but needs a lot of processing. The flowchart Fig. 2 shows the different stages of production from fiber to processing. The Environmental Protection Agency states that of the 15 pesticides used in cotton cultivation around seven are carcinogenic in nature and the US Agricultural Department states that in the year 2000 about 84 million pounds (38,101,759.08 kgs) of pesticides and 2 billion pounds (907,184,740 kgs) have been spread around 14.4 million acres (58,274.73 km^2) of farm land to produce cotton [4]. The Aral Sea in Central Asia, the Indus River in Pakistan, and the Murray Darling Basin in Australia are most affected by cotton cultivation. About 97% of the water from the Indus River goes for the cultivation of cotton leaving the river high and dry as seen in Fig. 3 [55]. Soil erosion and degradation, water contamination of lakes, rivers, wetlands, and other sources of water also tend to affect the biodiversity in and around the downstream areas.

Fabric has to be pretreated for proper absorption and fixation of the dyes or finishes. Dye penetration is hindered by starches and contaminants, so in the preparatory stage they are removed which accounts for 50% of the BOD in the effluent [6, 31]. The composition of the effluent is given in Table 1.

Starch, sodium chlorite, sodium bromite, and dilute mineral acids are retrieved from the pretreatment wastewater.Dyeing and finishing are important stages of textile manufacture which produces waste water. Dyeing effluent composition is given in Table 1. In the dyeing of fabrics, handling of toxic chemicals is required; acids like sulfuric and nitric acids are used for nitration with toluene and benzene [31]. The workers involved in the dyeing process are either directly or indirectly involved like skin burns in direct contact and respiratory problems (bronchitis and pulmonary edema) by nitric oxide or nitrous fumes due to indirect contact [15]. Denim is an

Fig. 3 River Indus [55]

Table 1 Average composition of textile dyeing waste water [55]

S. No.	Substances	Values
	pH	9.8–11.8
	Alkalinity	17–22 mg/l as $CaCO_3$
	BOD	760–900 mg/l
	COD	1400–1700 mg/l
	Total solids	6000–7000 mg/l
	Total chromium	10–13 mg/l

all-time favorite for many consumers, and there has been an array of finishes worked on denim—stone wash, acid wash, moon wash, monkey wash, frosted wash, mud wash, distressed wash, etc. The Levis Strauss published the life cycle assessment of its classic Levi's 501s style which reported that 920 gallons [3482.58 L] of water, 400 Mj of energy and 32 kgs of CO_2 was the requirement for manufacture of one pair of jeans. To make things easy to understand, this requirement was equivalent to using a garden hose for 106 min/ driving 78 miles/power for working 556 h with a computer as specified by Levis [2].

5 Design Strategies for Sustainable Products

Many studies and reports have shown that consumers are aware of the problems related to fast fashion and are looking out for sustainable products. Sustainable design formulation calls for assurance of a long-term product which satisfies the need of the customer. The ***function or use value*** of the product is of importance rather than the exchange value. In the case of textiles and apparels, it is very difficult for consumers to estimate the life of the garments; hence, manufacturers can provide additional information on the life span of the garment, the number of washes the garment will take and yet look good, the environmental benefits of the products. This will help the consumer to evaluate the connections between price, quality, and utility. A fashion product that aims for long life and deep satisfaction can be termed as ***slow design or slow fashion***. This product will be characterized by high quality, ethical values, and classic colors produced from 'age well' materials [20, 42]. When the customer has a deep sense of attachment with the product, the probability of using the product for a long time is inevitable. The designer must understand the longtime needs of the consumer and create products in tune with the consumer's needs—***co-creation*** [22]. This requires a new and innovative mind-set and new manufacturing strategies in all aspects of the manufacture of the product.

Mass customization is the production of tailor-made or custom-made products based on the individual needs of the customer at prices quoted by mass-produced products. This system integrates technologies like computerization, modularization, Internet, and lean production and creates an environment where every customer can have a product to his or her choice [11, 35, 44]. Customization helps the consumer to create personalized products which create a bonding between product and consumer. As online businesses address the fragmented markets across the globe in a personalized way, this innovation tends to satisfy both the manufacturer and individual user. Concept of ***half-way products*** provides a wide range of combinations based on the creativity of the individual. Products are available as kits and designed for disassembly giving the user the choice of new creations with the same materials, thereby adding a personal touch and memories to the individual, e.g., convertible clothing that has many parts which can be assembled to create the ensemble. Kit-based products help the user to develop new items and also get trained to repair the same. The next design strategy is the ***modular structures*** [20, 43] which help quick disassembly and reassembly of modules. The consumer can select details and trims as he/she wishes in terms of colors, materials, shapes, and silhouettes. Garments can be designed with certain functions in mind, e.g., collar and cuff alone can be detached and laundered as and when required [13].

In this design strategy ***co-design***, the end users are actively involved in the design process. When the stakeholder gets involved in the development process, there is a deep sense of satisfaction and develops an insight to understand and learn problems and solutions to develop a design. There is a sense of fulfillment when the end users complete the design with active involvement in the design process. ***Open-source fashion*** is a methodology where the designers market their design skills by supplying patterns and construction techniques as information to the consumer who implements the design information into a product. It is similar to the DIY information seen on the Internet resulting in immense satisfaction to the customer for the involvement taken to develop the product. ***Product service systems*** [13, 38, 47, 48] help to lower the environmental degradation as a set of products are used innumerable times and returned to the servicing organization. When customers invest in services and purchase functions, procuring new products, the use of virgin material and developing exercise for new product development will be reduced when compared to the traditional methods. This will facilitate the 'Zero-emission Society' where all

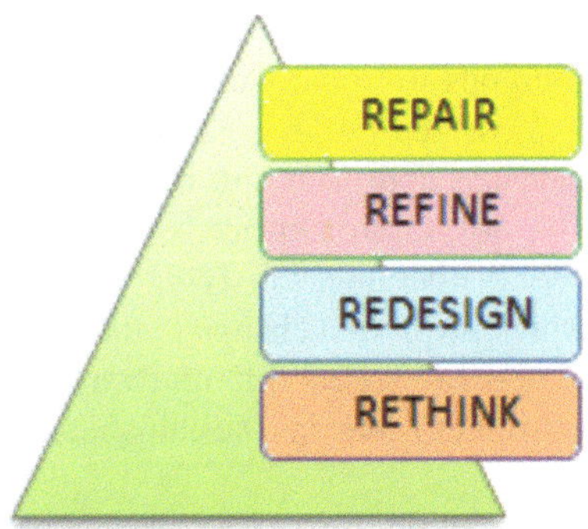

Fig. 4 Sustainable design strategies [52]

materials will be continuously returned to the producer after the use phase resulting in great environmental significance. Waste streams are avoided and the satisfaction of the customer is the key to services. Services also provide multiple options for the consumer to pick and choose offering a flexible product utilization which is not offered by real ownership of product.

Services for high and extensive utilization Consumers often think that it is better to buy a product when they see it as it may be handy one day, as it may useful when guests arrive, as it may decorate the home which we will build one day, as it may serve to be part of a huge collection. Products often purchased lie idle in wardrobes and in India most of them are passed on to somebody in the family whether they need it or not. Intensive utilization of products is achieved by renting or leasing, replacement of goods and services, shared use of goods, online garment exchange stocks, nonprofit networks for lending and sharing. Some accessories for ensembles like bows, ties, belts, umbrella, waistcoats, blazers, coats, hats, and scarves are not used frequently but can become rigorously utilized if they are shared with others through a renting system. These concepts highlight that an elemental change is essential for all business models to benefit the consumers. To have long-term use products should have classic designs, high quality, durability, and adjustable fit for more opportunities. Supplementary services will enable the extension of the life of the product by upgrading, updating, repairing, partial modification of the product or modular replacements. This creates a joy in using the product and also avoids the missed feeling that the customer experience once the product is no longer useable.

Sustainable design strategies, Fig. 4, include the 4 R's which help in reducing the environmental impact of products. The emphasis today is on ***repair*** of used products to extend their life by means of services made available to the consumers, e.g., if the size of a used apparel is not fitting a consumer, alterations can be made in the apparel by extending the seams and making use of the allowances given. Under the ***refine*** approach changes to existing products are made at the operational level to increase the eco-efficiency of the existing products. For example, a fabric may be dyed with reactive dye using caustic soda as auxiliary, this can be replaced by eco alkali [52] or bifunctional reactive dyes [14] can replace the uni-functional conventional reactive dyes, to reduce the environmental impact. This requires research and development to check if every stage is compatible. However, ***redesign*** approach calls for new solutions to obtain sustainable consumption coupled with environmental benefits to close the cycle, e.g., Fish skin as an alternative to animal leather requiring a redesign approach and removes the pollution associated with leather processing [39, 56]. Customer needs can be fulfilled by different approaches like something very demanding, satisfying yet environmentally safe. ***Rethinking*** is the next stage which involves a change in the mind-set of the consumer, and this change can bring in new life styles, different ways of living and doing things and also addresses the consumer needs in a sustainable manner. Statements like

> Beauty begins the moment you decide to be yourself
> -Coco Chanel
>
> Being true to yourself never goes out of style
> -Legally Blonde
>
> The only real elegance is in the mind; if you've got that, the rest really comes from it.
> -Diana Vreeland

If the attitude of the consumer is minimalistic, long lasting, and natural, then automatically the consumption of textile and apparels will be sustainable. Hence, rethinking enables this methodology and moves the philosophy toward sustainability. Fast fashion will be replaced by classical slow fashion; the consumer nod will be toward eco-friendly substitutes, e.g., Lakme Fashion week had the eco-friendly LOTUS cosmetic brand made available not only for the fashion artists but also for the onlookers, visitors, and participants. This hands-on experience provided an impetus to rethink and look out for natural eco-alternatives. New practices often hidden in the research laboratories should reach commercialization for providing new options and alternatives to pick and choose by the consumer.

To achieve sustainability in business models, a new outlook in value creation is required. Value is something about the product that makes the customer goes in for the purchase because they realize that is what the customer wants and it is better in comparison to the other products in the market. There are three value creation systems—***the core value*** (the basic value- reconceptualization) all manufacturing systems traditionally depend on this value; ***value-added*** (uses added innovations to reshape the existing) which includes services attached to the core value, e.g., customer involvement for use of modular products which needs updation and modifications from time to time; ***future-oriented value*** is usually associated with new business products and functions. This requires serious innovative design strategies, all-level collaboration and networking to ensure radical changes in the existing system with emphasis on sustainable production and consumption. The customer is the central point and customer needs and satisfaction is the starting point around which all opportunities evolve. The time has come for the current industrial economy to move on into a service-oriented functional economy with sustainability at its epicenter. Manufacturing was conducted on a global scale buying fabric from one country, accessories from different states, and production in another part of the world bearing in mind only cost-effectiveness with no concern on environment and society. The new vision is to use locally available materials for production to sustain the local culture, reducing the negative impact of logistics. This vision promotes users' needs and functions as primary aim which ends in a satisfying experience and emotional bonding with high-intrinsic product quality expanded by value-added services and longer life. Research and technology play important partners in executing the new visions of many business models. It is very difficult to convince today's customer as he/she is very informed and data as proof is very essential to make them work toward a particular direction—sustainability

6 Concluding Remarks

The apparel sector has been predicted to grow tremendously to cater to the huge population in India and China. The global middle class has grown from 3 billion in 2015 and will be 5.4 million by 2030 [28] leading to a huge demand. There will be a need for three times the natural resources used in 2000 if the consumption rate continues at the current rate. Apart from economic and social upheavals, there will be a large usage of virgin material which is finite and will require a long time to regenerate. The take-make-waste policy has to vanish and the apparel industries must produce less to implement slow fashion which requires a lot of reorientation and renovation of policies and practices. Only those business models that innovate and redesign with environmental concerns will be able to fit into the future markets.

To start the analysis to know where they stand all apparel manufacturers need to estimate their footprint on all three aspects of sustainability—economic, environmental, and social. The Higg Index, Sustainable Apparel Coalition offers a tool to measure the environmental, social and labor impacts of products and services with guidance on science based targets. Many apparel industries have submitted their results and the SAC can take up numeric and data to make effective comparisons and assessments. Many establishments are aware of the environmental risks and are ready to react—the first goal being to reduce the future footprint when compared to the previous one. This attitude will tend to keep the progress dynamic and move the traditional one to a sustainable one. Publicizing and reporting the good results will lead to more consumer awareness and prevent greenwashing practices.

Many retailers have started collecting old stuff for discounts and are entering into recycling and new product strategies. Others are working toward resource efficiency to reduce the use of virgin resources. Still others are moving into services—repair and reform, rentals and use of second-hand clothing. Initially, only a small percentage of products in the company were sold under the eco-friendly label, but the shift is to attain sustainability in all processes inside the establishment and also in the supply chain that creates the product.

The supplier and manufacturers set the trends, but today it is the need of the customer that is of primary concern. The success of the green products is dependent on the attributes (both environmental concerns and need based) by which it can compete with the nongreen products. After the Industrial Revolution, many waves of change have taken place; this phase of change seeks technological innovation, social issues and takes the lead to mold the minds of the consumers. The motto to be followed is 'why the consumer needs the product' rather than offering a product as per the fancies of the supplier and retailer. The focus should be on what the user wants as an outcome from the product. The manufacturer has to design the most sustainable product to accomplish the outcome. Rather than concentrating on how to produce and market their produce, the outcome-driven strategy is the best approach.

Fashion is a highly segmented business and one solution will not suit all business models. Many options are available for the designers to pick and choose and help to make the product sustainable in all aspects. Be it ***Basics First offering the world's first line of cradle-to-cradle gold level certified safe and compostable T-shirt*** or ***Natura Sewing Thread with cradle-to-cradle gold level certificate*** *or* ***Stella McCartney's Gold level certified Wool Yarn***—each organization has chosen a vision and redesigned their processes to make their product meet the requirements and standards of sustainability. Henceforth, the Fashion Business must learn from the past and innovate new solutions and perspectives as Fletcher [20] states 'it uses yesterday's thinking to cope with the conditions of tomorrow'. A new sustainable mind-set is slowly emerging in both the minds of the personnel involved in the Fashion Business and the consumers who will use the products. The time will soon come when all products produced and used are sustainable in nature and are found in abundance so that there will be no non-sustainable alternatives.

References

1. Acrook (2018) The 2018 Walmart and Sam's Club Miracle Balloon Campaign takes flight. https://seattlechildrens.childrensmiraclenetworkhospitals.org/the-2018-walmart-and-sams-club-miracle-balloon-campaign-takes-flight/. Accessed 5 Feb 2019
2. ADEME/AFNOR (2011) Environmental labelling, jeans product category rules. Environmental Protection Agency, Version 8.0
3. Allwood JM, Laursen SEM, de Rodriguez C, Bocken NMP (2006). Well dressed? University of Cambridge Institute of Manufacturing, Cambridge, UK
4. Asif AKMAH (2017) An overview of sustainability on apparel manufacturing industry in Bangladesh. SJEE 5:1–12. https://doi.org/10.11648/j.sjee.20170501.11
5. Ault N (2019) Gap, H & M under fire for female labor conditions in Asia. https://www.supplychaindive.com/news/HM-Gap-female-labor-rights-Asia/525334/. Accessed 4 Feb 2019
6. Babu BR, Parande AK, Raghu S, Kumar TP (2007) Cotton textile processing: waste generation and effluent treatment, textile technology. J Cotton Sci 11:141–153
7. Batra SH (1985) Other long vegetable fibers: abaca, banana, sisal henequen, flax, ramie, hemp, sunn and coir. In: Lewin M, Pearce EM (eds) Handbook of fiber science and technology, fiber chemistry, vol IV. Marcel Dekker, New York, pp 15–22
8. Belz F, Peattie K (2011) Sustainability marketing: a global perspective, 3rd edn. Wiley, West Sussex, UK
9. Bly S, Gwozdz W, Reisch LA (2013) Exit from the High Street: an exploratory study of sustainable fashion pioneers. http://scorai.org/wp-content/uploads/Bly-et-al.-finaljccsubmission.pdf. Accessed 5 Feb 2019
10. Brismer A (2019) Green Strategy. https://circularfashion.com/resources/actors-and-projects/. Accessed 2 Feb 2019

11. Business Dictionary (2019) Mass Customization. http://www.businessdictionary.com/definition/mass-customization.html. Accessed 2 Feb 2019
12. Butler S (2018) H & M factories in Myanmar employed 14 year old workers. https://www.theguardian.com/business/2016/aug/21/hm-factories-myanmar-employed-14-year-old-workers. Accessed 3 Feb 2019
13. Charter M, Clark T (2007) Sustainable innovation. Key conclusions from sustainable innovation conferences 2003–2006. The Centre for Sustainable Design. http://cfsd.org.uk/Sustainable%20Innovation/Sustainable_Innovation_report.pdf. Accessed 5 Feb 2019
14. Chavan RB (2001) Environment-friendly dyeing process. IJFTR 26:93–100
15. DOE (Department of Environment) (2008) Guide for assessment of effluent treatment plants in EMP/EIA reports for textile industries. Ministry of Environment and Forest, Bangladesh, pp A–22
16. eBay Giving Works (2014) eBay Giving Works—What it is, How it works and How you can benefit. https://community.ebay.com/t5/eBay-for-Business/eBay-Giving-Works-What-It-Is-How-It-Works-and-How-You-Can/ba-p/26163534. Accessed 3 Feb 2019
17. eBay Giving Works (2019) Wounded Warrier Project. https://www.woundedwarriorproject.org/donate/ebay-giving-works. Accessed 4 Feb 2019
18. Facing Finance (2019) H & M: Violations of Labor Rights in Uzbekistan, Bangladesh and Cambodia. http://www.facing-finance.org/en/database/cases/violation-of-labour-rights-by-hm-in-uzbekistan-bangladesh-and-cambodia/. Accessed 4 Feb 2019
19. Fashion Positive (2019) Case studies. https://www.fashionpositive.org/case-studies. Accessed 1 Feb 2019
20. Fletcher K (2008) Sustainable fashion & textiles: design journeys. Earthscan, London
21. Fuad-Luke A (2009) Design activism: beautiful strangeness for a sustainable world. Earthscan, London
22. Green Strategy (2017) Predictions on sustainability and Fashion for 2018 and beyond. http://www.greenstrategy.se/predictions-on-sustainability-and-fashion-for-2018-and-beyond-2/. Accessed 5 Feb 2019
23. Griffen E (2014) A first look at the communication theory—hierarchy of needs of Abraham Maslow, 9th edn. McGraw-hill Education, UK. https://www.afirstlook.com/docs/hierarchy.pdf
24. Griffiths J, Dirvanauskas G (2018) Frocky Horror Show https://www.thesun.co.uk/fabulous/6231893/met-gala-2018-red-carpet-worst-dressed/. Accessed 5 Feb 2019
25. H & M Group (2018) H & M Conscious Exclusive 2018 brings together powerfulfeminity and sustainable fashion innovation with recycled silver and Econyl. https://about.hm.com/en/media/news/general-news-2018/hm-conscious-exclusive-2018-brings-together-powerful-femininity-and-sustainable-fashion-innovation-with-recycled-silver-and-econyl.html
26. H & M Group (2019) H & M Debuts Conscious Exclusive Collection from Autumn/Winter—Introducing Recycled Cashmere and Velvet made from Recycled Polyester. https://about.hm.com/en/media/news/general-news-2018/HM-debuts-Conscious-Exclusive-collection-for-Autumn-Winter-introducing-recycled-cashmere-and-velvet-made-from-recycled-polyester.html. Accessed 5 Feb 2019
27. Hegarthy S (2012) How jeans conquered the world. https://www.bbc.com/news/magazine-17101768. Accessed 2 Feb 2019
28. Hermes J (2017) Opinion: apparel Industry must embrace a new sustainable approach to meet demand. https://www.environmentalleader.com/2017/07/opinion-apparel-industry-must-embrace-new-sustainable-approach-meet-demand/. Accessed 6 Feb 2019
29. Hitchings-Hales J (2018) Girls & Women: Hundreds of H & M and Gap Factory Workers Abused Daily, Report Says. https://www.globalcitizen.org/en/content/hm-gap-factory-abuse-fast-fashion-workers/. Accessed 4 Feb 2019
30. Huitt W (2007) Maslow's hierarchy of needs. Educational Psychology Interactive, Valdosta, Valdosta University. https://www.iccb.org/iccb/wpcontent/pdfs/adulted/healthcare_curriculum/curriculum&resources/context_social_studies/F.%20HC%20Context%20Social%20Studies%20Resource%20File/Maslow%27s%20Heirarchy%20of%20Needs.pdf. Accessed 2 Feb 2019

31. Islam MM, Mahmud K, Faruk O, Billah MS (2011) Textile dyeing industries in Bangladesh for sustainable development. IJEST 2:428–436
32. Jackson T, Shaw D (2009) Mastering fashion marketing. Palgrave Macmillan, New York, USA
33. Kaikobad NK, Bhuiyan MZA, Zobaida HN, Daizy AH (2015) Sustainable and ethical fashion: the environmental and morality issues. IOSR-JHSS 20:17–22
34. Kell G (2018) Can Fashion be Sustainable. https://www.forbes.com/sites/georgkell/2018/06/04/can-fashion-be-sustainable/#369c85d0412b. Accessed 3 Feb 2019
35. Lee S, Chen J (2000) Mass-customization methodology for an apparel industry with a future. JIT 16:1–8
36. Martin D, Joomis K (2007) Building teachers: a constructive approach to introducing education. https://www.cengage.com/resource_uploads/downloads/0495570540_162121.pdf. Accessed 3 Feb 2019
37. McLeod (2018) Maslow's Hierarchy of Needs. https://www.simplypsychology.org/maslow.html. Accessed 4 Feb 2019
38. Mont O (2002) Clarifying the concept of product-service system. J Clean Prod 10:237–245
39. Moskvitch K (2018) Forget leather, the future of fashion is all about fish skin. https://www.wired.co.uk/article/fish-leather-shirts-sustainability-leather-demand. Accessed 6 Feb 2019
40. Niinimäki K (2011) From disposable to sustainable: the complex interplay between design and consumption of textiles and clothing. Doctoral dissertation, Aalto University, Helsinki. https://aaltodoc.aalto.fi/handle/123456789/13770. Accessed 1 Feb 2019
41. Niinimäki K (2014) Sustainable consumer satisfaction in the context of clothing. In: Vezzoli C, Kohtala C, Srinivasan A (eds) Product-service system design for sustainability. Greenleaf, Sheffield, UK, pp 218–237
42. Niinimäki K (2009) Developing sustainable products by deepening consumers' product attachment through customizing. In: Proceedings of the world conference on mass customization & personalization MCPC 2009. Helsinki, Finland
43. Papanek V (1995) The green imperative: ecology and ethics in design and architecture. Thames and Hudson, London
44. Pine J (1993) Mass customization. Harvard Business School Press, Boston
45. Pinterest (2019) 73 best compassion consciousness. https://www.google.com/search?tbm=isch&sa=1&ei=BpiEXMyfHaHfz7sPspeT8Aw&q=compassionate+fashion+%2B+quotes&oq=compassionate+fashion+%2B+quotes&gs_l=img.12…0.0..11575…0.0..0.0.0…….0……gws-wiz-img.RGROMvSFDR4. Accessed 3 Feb 2019
46. Radhakrishnan S (2017) Denim recycling. Textiles and clothing sustainability. Springer, Singapore, pp 79–125
47. Robert KH, Schmidt-Bleek B, de Larderel JA, Basile G, Jansen JL, Kuehr R, Thomas PP, Suzuki M, Hawken P, Wackernagel M (2002) Strategic sustainable development e selection, design and synergies of applied tools. J Clean Prod 10:197–214
48. Stahel WR (2001) Sustainability and services. In: Charter M, Tischner U (eds) Sustainable solutions: developing products and services for the future. Greenleaf, Sheffield
49. Statistica (2019) Womens Life style and fashion magazines ranked by print retail sales volume in the United Kingdom (UK) in the first half 2018 (in copies sold). https://www.statista.com/statistics/321619/women-s-lifestyle-magazines-ranked-by-sales-volume-uk/. Accessed 3 Feb 2019
50. Textiles Environmental Design (2019) TED RESEARCH—The Ten. https://circularfashion.com/resources/actors-and-projects/. Accessed 3 Feb 2019
51. The Children's Hospital of San Antonio Foundation (2019) Walmart and Sam's Club Associates making miracles happen. https://www.childrenshospitalsafoundation.org/walmart-and-sams-club-associates-making-miracleshappen/. Accessed 5 Feb 2019
52. Uddin GM, Gosh NC, Reza MS (2014) Study on the performance of Eco-Alkali in dyeing of cotton fabric with reactive dyes. Int J Text Sci 3:51–58
53. UK Essays (2017) Consumer Behaviour in the Fashion Market. https://www.ukessays.com/essays/marketing/consumer-behaviour-in-the-fashion-market-marketing-essay.php. Accessed 3 Feb 2019

54. Venugopalan O (2007) Maslow's theory of motivation its relevance and application among non-managerial employees of selected public and private sector undertakings in Kerala. Dissertation, University of Calicut
55. World Wildlife Fund (2019) Sustainable Agriculture- Cotton-impacts. https://www.worldwildlife.org/industries/cotton. Accessed 5 Feb 2019
56. Zengin ACA, Basaran B, Karvana HA, Mutlu MM, Bitlisli BO, Gaidau C, Niculeicu M, Maureanu M (2015) Fish skins: valuable resources for leather industry. XXXIII IULTCS Congress. Novo Hamburgo/Brazil, pp 24–27, Nov 2015

Zeitfracht Medien GmbH
Ferdinand-Jühlke-Straße 7
99095 Erfurt, Deutschland
produktsicherheit@kolibri360.de